A Unified Theory of Nonlinear Operator and Evolution Equations with Applications

PURE AND APPLIED MATHEMATICS

A Program of Monographs, Textbooks, and Lecture Notes

MONOGRAPHS AND TEXTBOOKS IN PURE AND APPLIED MATHEMATICS

1. *K. Yano,* Integral Formulas in Riemannian Geometry (1970)*(out of print)*
2. *S. Kobayashi,* Hyperbolic Manifolds and Holomorphic Mappings (1970) *(out of print)*
3. *V. S. Vladimirov,* Equations of Mathematical Physics (A. Jeffrey, editor; A. Littlewood, translator) (1970) *(out of print)*
4. *B. N. Pshenichnyi,* Necessary Conditions for an Extremum (L. Neustadt, translation editor; K. Makowski, translator) (1971)
5. *L. Narici, E. Beckenstein, and G. Bachman,* Functional Analysis and Valuation Theory (1971)
6. *D. S. Passman,* Infinite Group Rings (1971)
7. *L. Dornhoff,* Group Representation Theory (in two parts). Part A: Ordinary Representation Theory. Part B: Modular Representation Theory (1971, 1972)
8. *W. Boothby and G. L. Weiss (eds.),* Symmetric Spaces: Short Courses Presented at Washington University (1972)
9. *Y. Matsushima,* Differentiable Manifolds (E. T. Kobayashi, translator) (1972)
10. *L. E. Ward, Jr.,* Topology: An Outline for a First Course (1972) *(out of print)*
11. *A. Babakhanian,* Cohomological Methods in Group Theory (1972)
12. *R. Gilmer,* Multiplicative Ideal Theory (1972)
13. *J. Yeh,* Stochastic Processes and the Wiener Integral (1973) *(out of print)*
14. *J. Barros-Neto,* Introduction to the Theory of Distributions (1973) *(out of print)*
15. *R. Larsen,* Functional Analysis: An Introduction (1973) *(out of print)*
16. *K. Yano and S. Ishihara,* Tangent and Cotangent Bundles: Differential Geometry (1973) *(out of print)*
17. *C. Procesi,* Rings with Polynomial Identities (1973)
18. *R. Hermann,* Geometry, Physics, and Systems (1973)
19. *N. R. Wallach,* Harmonic Analysis on Homogeneous Spaces (1973) *(out of print)*
20. *J. Dieudonné,* Introduction to the Theory of Formal Groups (1973)
21. *I. Vaisman,* Cohomology and Differential Forms (1973)
22. *B. -Y. Chen,* Geometry of Submanifolds (1973)
23. *M. Marcus,* Finite Dimensional Multilinear Algebra (in two parts) (1973, 1975)
24. *R. Larsen,* Banach Algebras: An Introduction (1973)
25. *R. O. Kujala and A. L. Vitter (eds.),* Value Distribution Theory: Part A; Part B: Deficit and Bezout Estimates by Wilhelm Stoll (1973)
26. *K. B. Stolarsky,* Algebraic Numbers and Diophantine Approximation (1974)
27. *A. R. Magid,* The Separable Galois Theory of Commutative Rings (1974)
28. *B. R. McDonald,* Finite Rings with Identity (1974)
29. *J. Satake,* Linear Algebra (S. Koh, T. A. Akiba, and S. Ihara, translators) (1975)

30. *J. S. Golan,* Localization of Noncommutative Rings (1975)
31. *G. Klambauer,* Mathematical Analysis (1975)
32. *M. K. Agoston,* Algebraic Topology: A First Course (1976)
33. *K. R. Goodearl,* Ring Theory: Nonsingular Rings and Modules (1976)
34. *L. E. Mansfield,* Linear Algebra with Geometric Applications: Selected Topics (1976)
35. *N. J. Pullman,* Matrix Theory and Its Applications (1976)
36. *B. R. McDonald,* Geometric Algebra Over Local Rings (1976)
37. *C. W. Groetsch,* Generalized Inverses of Linear Operators: Representation and Approximation (1977)
38. *J. E. Kuczkowski and J. L. Gersting,* Abstract Algebra: A First Look (1977)
39. *C. O. Christenson and W. L. Voxman,* Aspects of Topology (1977)
40. *M. Nagata,* Field Theory (1977)
41. *R. L. Long,* Algebraic Number Theory (1977)
42. *W. F. Pfeffer,* Integrals and Measures (1977)
43. *R. L. Wheeden and A. Zygmund,* Measure and Integral: An Introduction to Real Analysis (1977)
44. *J. H. Curtiss,* Introduction to Functions of a Complex Variable (1978)
45. *K. Hrbacek and T. Jech,* Introduction to Set Theory (1978)
46. *W. S. Massey,* Homology and Cohomology Theory (1978)
47. *M. Marcus,* Introduction to Modern Algebra (1978)
48. *E. C. Young,* Vector and Tensor Analysis (1978)
49. *S. B. Nadler, Jr.,* Hyperspaces of Sets (1978)
50. *S. K. Segal,* Topics in Group Rings (1978)
51. *A. C. M. van Rooij,* Non-Archimedean Functional Analysis (1978)
52. *L. Corwin and R. Szczarba,* Calculus in Vector Spaces (1979)
53. *C. Sadosky,* Interpolation of Operators and Singular Integrals: An Introduction to Harmonic Analysis (1979)
54. *J. Cronin,* Differential Equations: Introduction and Quantitative Theory (1980)
55. *C. W. Groetsch,* Elements of Applicable Functional Analysis (1980)
56. *I. Vaisman,* Foundations of Three-Dimensional Euclidean Geometry (1980)
57. *H. I. Freedman,* Deterministic Mathematical Models in Population Ecology (1980)
58. *S. B. Chae,* Lebesgue Integration (1980)
59. *C. S. Rees, S. M. Shah, and C. V. Stanojević,* Theory and Applications of Fourier Analysis (1981)
60. *L. Nachbin,* Introduction to Functional Analysis: Banach Spaces and Differential Calculus (R. M. Aron, translator) (1981)
61. *G. Orzech and M. Orzech,* Plane Algebraic Curves: An Introduction Via Valuations (1981)
62. *R. Johnsonbaugh and W. E. Pfaffenberger,* Foundations of Mathematical Analysis (1981)
63. *W. L. Voxman and R. H. Goetschel,* Advanced Calculus: An Introduction to Modern Analysis (1981)
64. *L. J. Corwin and R. H. Szcarba,* Multivariable Calculus (1982)
65. *V. I. Istrătescu,* Introduction to Linear Operator Theory (1981)
66. *R. D. Järvinen,* Finite and Infinite Dimensional Linear Spaces: A Comparative Study in Algebraic and Analytic Settings (1981)

67. *J. K. Beem and P. E. Ehrlich*, Global Lorentzian Geometry (1981)
68. *D. L. Armacost*, The Structure of Locally Compact Abelian Groups (1981)
69. *J. W. Brewer and M. K. Smith, eds.*, Emmy Noether: A Tribute to Her Life and Work (1981)
70. *K. H. Kim*, Boolean Matrix Theory and Applications (1982)
71. *T. W. Wieting*, The Mathematical Theory of Chromatic Plane Ornaments (1982)
72. *D. B. Gauld*, Differential Topology: An Introduction (1982)
73. *R. L. Faber*, Foundations of Euclidean and Non-Euclidean Geometry (1983)
74. *M. Carmeli*, Statistical Theory and Random Matrices (1983)
75. *J. H. Carruth, J. A. Hildebrant, and R. J. Koch*, The Theory of Topological Semigroups (1983)
76. *R. L. Faber*, Differential Geometry and Relativity Theory: An Introduction (1983)
77. *S. Barnett*, Polynomials and Linear Control Systems (1983)
78. *G. Karpilovsky*, Commutative Group Algebras (1983)
79. *F. Van Oystacyen and A. Verschoren*, Relative Invariants of Rings: The Commutative Theory (1983)
80. *I. Vaisman*, A First Course in Differential Geometry (1984)
81. *G. W. Swan*, Applications of Optimal Control Theory in Biomedicine (1984)
82. *T. Petrie and J. D. Randall*, Transformation Groups on Manifolds (1984)
83. *K. Goebel and S. Reich*, Uniform Convexity, Hyperbolic Geometry, and Nonexpansive Mappings (1984)
84. *T. Albu and C. Năstăsescu*, Relative Finiteness in Module Theory (1984)
85. *K. Hrbacek and T. Jech*, Introduction to Set Theory, Second Edition, Revised and Expanded (1984)
86. *F. Van Oystaeyen and A. Verschoren*, Relative Invariants of Rings: The Noncommutative Theory (1984)
87. *B. R. McDonald*, Linear Algebra Over Commutative Rings (1984)
88. *M. Namba*, Geometry of Projective Algebraic Curves (1984)
89. *G. F. Webb*, Theory of Nonlinear Age-Dependent Population Dynamics (1985)
90. *M. R. Bremner, R. V. Moody, and J. Patera*, Tables of Dominant Weight Multiplicities for Representations of Simple Lie Algebras (1985)
91. *A. E. Fekete*, Real Linear Algebra (1985)
92. *S. B. Chae*, Holomorphy and Calculus in Normed Spaces (1985)
93. *A. J. Jerri*, Introduction to Integral Equations with Applications (1985)
94. *G. Karpilovsky*, Projective Representations of Finite Groups (1985)
95. *L. Narici and E. Beckenstein*, Topological Vector Spaces (1985)
96. *J. Weeks*, The Shape of Space: How to Visualize Surfaces and Three-Dimensional Manifolds (1985)
97. *P. R. Gribik and K. O. Kortanek*, Extremal Methods of Operations Research (1985)
98. *J.-A. Chao and W. A. Woyczynski, eds.*, Probability Theory and Harmonic Analysis (1986)
99. *G. D. Crown, M. H. Fenrick, and R. J. Valenza*, Abstract Algebra (1986)
100. *J. H. Carruth, J. A. Hildebrant, and R. J. Koch*, The Theory of Topological Semigroups, Volume 2 (1986)

101. *R. S. Doran and V. A. Belfi,* Characterizations of C*-Algebras: The Gelfand-Naimark Theorems (1986)

102. *M. W. Jeter,* Mathematical Programming: An Introduction to Optimization (1986)

103. *M. Altman*, A Unified Theory of Nonlinear Operator and Evolution Equations with Applications: A New Approach to Nonlinear Partial Differential Equations (1986)

Other Volumes in Preparation

A Unified Theory of Nonlinear Operator and Evolution Equations with Applications

A NEW APPROACH TO NONLINEAR PARTIAL DIFFERENTIAL EQUATIONS

Mieczyslaw Altman
Louisiana State University
Baton Rouge, Louisiana

MARCEL DEKKER, INC. New York and Basel

Library of Congress Cataloging-in-Publication Data

Altman, Mieczyslaw, [date]
A unified theory of nonlinear operator and evolution equations with applications.

(Monographs and textbooks in pure and applied mathematics ; 103)
Bibliography ; p.
Includes index.
1. Operator equations, Nonlinear, 2. Evolution equations, Nonlinear. I. Title. II. Series.
QA329.8.A57 1986 515.7'248 86-9042
ISBN 0-8247-7613-5

MARCEL DEKKER, INC.
270 Madison Avenue, New York, New York 10016

Current printing (last digit):
10 9 8 7 6 5 4 3 2 1

PRINTED IN THE UNITED STATES OF AMERICA

Life is not choosing between good and evil.
Life is a struggle for survival or for a better life.
That "better" tells a human being from any other creature.

The victims of the Holocaust lost their struggle for survival.
To Them, including my Parents, Brothers and Sisters, Relatives and Friends, I dedicate this work.

Preface

This monograph presents a unified approach to both nonlinear operator and evolution equations. One rather important feature of our theory of nonlinear evolution equations in Banach spaces is that nonlinear problems can be solved via linearized evolution equations. From the standpoint of applications to nonlinear partial differential equations this means that such equations can be solved if the linearized problem can be treated as a linear evolution equation, based on the C_0-semigroups theory.

Our work is an outgrowth of research that started in 1980. In 1981 Kato's famous theorem on quasilinear evolution equations was generalized to nonreflexive Banach spaces by means of a method of contractor directions. The next step was aimed at nonlinear evolution equations. In order to handle the "loss of derivatives," in 1956 Nash invented the concept of smoothing operators, and in 1966 Moser introduced the concept of the degree of approximate linearization. Neither Nash's nor Moser's theorems applied to nonlinear abstract evolution equations.* New concepts were needed, for example, convex approximate linearization and the degree of elliptic regularization. The best results followed from a combination of smoothing operators with elliptic regularization. After the development of the theory for nonlinear evolution equations it turned out that a

*See Remark 3.2, Chapter 6.

parallel theory holds true for nonlinear operator equations, thus yielding a unified approach to both theories.

The term *elliptic regularization* is used in a broader sense, for in the case of smoothing operators no "gain of derivatives" is needed in the a priori estimates of the solutions of the linearized equations. Moreover, a new notion of the degree of approximate linearization is indicated which can be applied to both nonlinear operator and evolution equations. Finally, the theory based on GLIM-I and II (described below) can also be applied to optimization and optimal control problems with "loss of derivatives," in particular, to optimal control of systems governed by abstract nonlinear evolution equations. The very nature of these problems requires an approximate linearization approach.

This monograph presents a unified theory that makes a further contribution to the Nash-Moser technique of solving nonlinear operator equations. It also provides a new theory of nonlinear evolution equations, based mainly on the possibility of solving nonlinear evolution equations via linearized evolution equations. In this way the powerful theory of linear evolution equations can be utilized to solve nonlinear evolution equations. This fact is rather important from the standpoint of applications to nonlinear partial differential equations. Moreover, both theories have the advantage that no "gains of derivatives" are needed in the a priori estimates of the solutions of the linearized equations.

The first six chapters deal with nonlinear operator equations, while Chapters 7 through 11 are devoted to evolution equations. The operator theory section consists of three parts: the first contains a general existence and convergence scheme, the second combines smoothing operators with elliptic regularization, and the third uses elliptic regularization only. Each part is treated by three different methods. The same scheme is followed for nonlinear evolution equations. In the case of strongly quasilinear evolution equations the only applicable method is the method of contractor directors.

Since the publication of *Contractors and Contractor Directions: Theory and Applications* (Altman, 1977), considerable progress has been made in the area. One of the most important facts has been the development of a theory of iterative methods of contractor directions, which permitted Kato's theorem on quasilinear evolution equations to be generalized to nonreflexive Banach spaces. But for a nonlinear theory of evolution equations this was insufficient, as was the Nash-Moser technique. The new concepts needed for this purpose are discussed in Chapters 7 and 8. Most methods for solving nonlinear problems are based on the essential fact of *local* linearization, which means that the nonlinear equations are linearized about a vector running over a bounded subset B. This holds for topological methods (for example, the Schauder fixed point theorem and its generalizations, and the Leray-Schauder degree and its generalizations) and also for iterative methods [for example, the Banach contraction principle, the Newton-Kantorovič method, and methods of contractors and contractor directors (Altman, 1977; 1980)]. Another important example is Kato's (1975) theory of quasilinear evolution equations, which he successfully applied to partial differential equations of mathematical physics. Kato's theory is applicable only to reflexive Banach spaces. A generalization to nonreflexive Banach spaces is given in Altman (1981; 1983a,b). But the method used there is different. It is based on the notion of contractor directions (see Altman, 1983a,b) and it is independent of the C_0-semigroups theory on which Kato's technique relies completely. Both quasilinear theories involve iterative methods based on local linearization.

However, there are some nonlinear problems, involving "loss of derivatives," that are intractable by the above methods. Nash (1956) introduced a procedure for attacking such problems, and Moser (1966) introduced a scheme based on the Newton-Kantorovič method. Another modification of the Nash method was given by Hörmander (1976).

These methods can be characterized as iterative methods based on *global* (i.e., B is unbounded) linearization in contrast to the ones based on *local* linearization mentioned above. Nash used a one-parameter family of linear smoothing operators to obtain smooth approximate solutions of the linearized equations, while Moser's technique was based on the notion of the degree of the approximate linearization.

The new idea behind the Nash-Moser technique was the motivation for our attempt to solve the problem of existence of solutions to nonlinear evolution equations. It should be emphasized that the Nash-Moser technique is designed for nonlinear operator equations and the methods mentioned above are not applicable to nonlinear evolution equations. In a series of investigations to solve the latter problem we present three different global linearization iterative methods:

GLIM-I is an iterative method with small steps which is based on the concept of contractor directions.

GLIM-II is an iterative method with stepsize 1 which is independent of the notion of contractor directions.

GLIM-III is a rapidly convergent iteration method which is based on a further development of Moser's essential ideas.

Three new concepts are involved in the above methods: convex approximate linearization, the related notion of convex linearization implicitly employed in Altman (1981; 1983a,b), and the order of approximate linearization, which generalizes Moser's notion of the degree of the same. As far as the construction of smooth approximate solutions of the linearized equations is concerned, we use elliptic regularization or smoothing operators combined with elliptic regularization involving the degree of elliptic regularization.

The theory of C_0-semigroups is in general crucial to the problem of finding smooth approximate solutions of the linearized equations, because of the possibility of solving nonlinear evolution equations via linearized evolution equations. A number of results have been obtained by using the theory of nonlinear semigroups to obtain existence theorems of solutions for nonlinear evolution equations. (For references, see Barbu, 1976.) But the nonlinear theory is, of course, not as powerful as the C_0-semigroups theory.

Our approach has one more advantage: Fortunately, the three methods mentioned above also apply to general nonlinear operator equations in Banach spaces. Moreover, it is possible to consider the more general class of operators $P = L + F + f$, where L is a linear unbounded operator, F is strictly nonlinear, and f is a (semilinear) locally Lipschitzian term. However, it appears that GLIM-I is the most powerful and flexible method. Needless to say, none of the methods available in the literature is applicable to the class of P determined above.

An extensive presentation of the alternative or bifurcation method for solving the equation $Lx = Fx$ is given by Cesari (1976), based on topological arguments, Schauder's and Banach's fixed point theorems, and monotone operators. But the reasoning used there depends on local linearization and therefore is not applicable to general problems with "loss of derivatives." Based on the method of finite differences, a number of interesting results for nonlinear and quasilinear evolution equations have been obtained by Hazan (1973; 1982; 1983); see also Kobayashi (1983) and the references therein.

This monograph is a recapitulation of our published, submitted, and unpublished results obtained since 1980. While the book contains no background material, the reader familiar with a first course in functional analysis will be able to grasp the main thrust of the text. I owe special thanks to Marcel Dekker, Inc., for making available this new development in mathematical research. We hope that our unified theory will open new directions and challenges in both theory and applications.

Mieczyslaw Altman

Contents

1
Convex Approximate Linearization and Global Linearization Iterative Methods

A general scheme for solving the nonlinear operator equation (1.1) is presented in this chapter, where each of the three sections contains a different method. The idea behind *convex* approximate linearization can be explained as follows. Since the operator P in (1.1) contains an unbounded linear operator L, the "usual" linearized equation is replaced by (1.2), where the approximate solution z can be estimated in terms of x. Then by putting $x = z + h$, it follows that h satisfies the authentic linearized equation (1.10). Three spaces $Z \subset Y \subset X$ are needed so that $Px_n \to 0$ in the base space X, where $x_n = x_0 + h_1 + \cdots + h_{n-1}$ belong to Z and $\|x_n\|_Z < K_n$ with $K_n \to \infty$. By the interpolation property (A_0) the sequence $\{x_n\}$ converges in Y to a solution of (1.1). Moser's (1966) theorem is not applicable to (1.1) because it requires $L \equiv 0$ and $f \equiv 0$. Besides, his condition that $\|h\|_0 \le C\|F'(x)h\|$ is replaced by (1.8), which is a weaker one. The methods presented in this chapter do not indicate how to obtain approximate solutions of the linearized equations; neither does Moser's (1966) method. This will be accomplished in Chapters 2 through 6.

1. AN ITERATIVE METHOD OF CONTRACTOR DIRECTIONS, GLIM-I

Let $Z \subset Y \subset X$ be Banach spaces with norms $\|\cdot\|_Z \ge \|\cdot\|_Y \ge \|\cdot\|_X$. We assume that there exist positive constants C, $\bar{s}$ with $0 < \bar{s} < 1$ such that

(A_0) $\|x\|_Y \le C\|x\|_X^{1-\bar{s}}\|x\|_Z^{\bar{s}}$

Let W_0 be an open ball in Y with center x_0 in Z and radius $r > 0$. Put $V_0 = W_0 \cap Z$ and let V_1 be the $\|\cdot\|_Y$-closure of V_0.

Let $P \equiv L + F + f$ be a mapping $P\colon V_1 \to E$, another Banach space with norm $\|\cdot\|$, or $E = X$, where L is a nonbounded linear operator, F being strictly nonlinear, and f locally Lipschitzian. If the domain of P is $\mathcal{D}$, then we put $V_0 = \mathcal{D} \cap W_0 \cap Z$. Consider the nonlinear operator equation

$$Px \equiv Lx + Fx + fx = 0 \tag{1.1}$$

We assume that F is differentiable; i.e., for each $x \in V_0$, there exists a linear operator $F'(x)$ such that

$$\varepsilon^{-1}\|F(x + \varepsilon h) - Fx - \varepsilon F'(x)h\| \to 0$$

as $\varepsilon \to 0+$, where $h \in Y$.

1.1 DEFINITION (Notation) Let $\mu > 0$, $0 \le \nu < 1$, $\sigma \ge 0$ be given numbers. Then the linearized equation

$$Lz + F'(x)z + Fx - F'(x)x + fx = 0 \tag{1.2}$$

admits approximate solutions of order (μ,ν,σ) if there exists a constant $M > 0$ with the following property. For every $x \in V_0$, $K > 1$, and $Q > 1$, if $\|x\|_Z < K$ then there exist a residual (error) vector y and a vector z such that

$$\|z\|_Z \le MQK^{\nu} \tag{1.3}$$

$$\|y\| \le MQ^{-\mu}K^{\sigma} \tag{1.4}$$

and

$$Lz + F'(x)z + Fx - F'(x)x + fx + y = 0 \tag{1.5}$$

We make the following assumptions.

(A_1) Let $\{x_n\} \subset V_0$ be a Cauchy sequence in Y and let $\{h_n\} \subset Z$ be bounded in Y. Then $\varepsilon_n \to 0$ implies

$$\varepsilon_n^{-1}\|F(x_n + \varepsilon_n h_n) - Fx_n - \varepsilon_n F'(x_n)h_n\| \to 0 \text{ as } n \to \infty \tag{1.6}$$

P is closed on V_1.

(A_2) There exists a constant $C_0 > 0$ with the following property. For $x \in V_0$ and $g \in E$, if h is a solution of the equation

$$Lh + F'(x)h + g = 0 \tag{1.7}$$

then

$$\|h\|_X \le C_0\|g\| \tag{1.8}$$

(A_3) There exists a constant $q_0 > 0$ such that

$$\|f(x + \varepsilon h) - fx\| \le q_0\varepsilon\|h\|_X \quad \text{and} \quad q_0C_0 < 1 \tag{1.9}$$

where $x \in V_0$, $x + \varepsilon h \in V_0$, $0 < \varepsilon \le 1$.

Now for $x \in V_0$, let z be a solution of the equation (1.5) and put $z = x + h$. Then obviously h is a solution of the equation

$$Lh + F'(x)h + Px + y = 0 \tag{1.10}$$

and we get from (1.3), (1.4), and (1.7) with g replaced by $Px + y$ that

$$\|h\|_Z \le \|x\|_Z + MQK^{\nu} \tag{1.11}$$

and

$$\|h\|_X \le C_0(\|Px\| + MQ^{-\mu}K^{\sigma}) \tag{1.12}$$

1.1 LEMMA Let $x \in V_0$ with $\|x\|_Z < K$, and let

$$q_0C_0 < q < 1 \text{ and } \bar{q} = (q - q_0C_0)/2 < q \tag{1.13}$$

If h is a solution of the equation (1.10), then there exists $0 < \varepsilon \le 1$ such that

$$\|P(x + \varepsilon h) - (1 - \varepsilon)Px\| \le \varepsilon q\|Px\| \tag{1.14}$$

and

$$\|h\|_X \le C_0(1 + \bar{q})\|Px\| \tag{1.15}$$

Proof: We use the following identity

$$P(x + \varepsilon h) - (1 - \varepsilon)Px = [F(x + \varepsilon h) - Fx - \varepsilon F'(x)h] + [f(x + \varepsilon h) - f(x)] + \varepsilon[Lh + F'(x)h + Px + y] - \varepsilon y$$

Hence, we obtain from (1.7), (1.12), (1.4),

$$\begin{aligned}\|P(x + \varepsilon h) - (1 - \varepsilon)Px\| &\le \|F(x + \varepsilon h) - Fx - \varepsilon F'(x)h\| + q_0\varepsilon\|h\|_X + \varepsilon\|y\| \\ &\le \|F(x + \varepsilon h) - Fx - \varepsilon F'(x)h\| + \varepsilon(q_0C_0\|Px\| + 2MQ^{-\mu}K^{\sigma})\end{aligned}$$

Hence

$$\|P(x + \varepsilon h) - (1 - \varepsilon)Px\| \le \|F(x + \varepsilon h) - Fx - \varepsilon F'(x)h\| + \varepsilon\bar{q}\|Px\| \tag{1.14a}$$

by (1.13), provided $Q > 1$ is such that

$$MQ^{-\mu}K^{\sigma} \le \bar{q}\|Px\| \tag{1.16}$$

whence relation (1.14) follows if ε is so small that

$$\|F(x + \varepsilon h) - Fx - \varepsilon F'(x)h\| < \varepsilon(q - \bar{q})\|Px\|$$

Relation (1.15) results from (1.16) and (1.12). ∎

1.2 LEMMA Let $x_n \in V_0$ and let

$$\|x_n\|_Z < A\exp(\alpha(1 - q)t_n) = K_n \tag{1.17}$$

where $t_n > 0$, and

$$\mu(1 - \nu) - \sigma > 0 \tag{1.18}$$

where $\alpha > 1$ and $A > 0$ are such that

$$\alpha(1 - q) - 1 > 0 \quad \text{and} \quad \alpha > [\mu(1 - \nu) - \sigma]^{-1} \tag{1.19}$$

and

$$MM^{1/\mu}(\bar{q}p_0)^{-1/\mu} < [\alpha(1 - q) - 1]A^{1-\nu-\sigma/\mu} \tag{1.20}$$

where $\|Px_0\| = p_0$.

Then there exists a number $Q = Q_n$ such that

$$MQ_n^{-\mu}K_n^{\sigma} < \bar{q}p_0 \exp(-(1 - q)t_n) \tag{1.21}$$

and

$$\|x_n + \varepsilon h_n\|_Z \le A \exp(\alpha(1 - q)(t_n + \varepsilon)) \tag{1.22}$$

where h_n is a solution of the equation (1.10) with $x = x_n$, $y = y_n$, satisfying relation (1.12) with $x = x_n$, $K = K_n$, and $Q = Q_n$.

Proof: Relation (1.11) with $x = x_n$ implies

$$\begin{aligned} \|x_n + \varepsilon h_n\|_Z &\le \|x_n\|_Z + \varepsilon\|h_n\|_Z \\ &\le (1 + \varepsilon)\|x_n\|_Z + \varepsilon MQ_n K_n^{\nu} \end{aligned}$$

Hence, relation (1.22) follows if

$$(1 + \varepsilon)A \exp(\alpha(1 - q)t_n) + \varepsilon MQK_n^{\nu} < A \exp(\alpha(1 - q)(t_n + \varepsilon)) \tag{1.23}$$

But it is easily seen that

$$\alpha(1 - q) - 1 = \min_{0\le\varepsilon\le 1} [\exp(\alpha(1 - q)\varepsilon) - (1 + \varepsilon)]/\varepsilon \tag{1.24}$$

Hence, by (1.17), relation (1.23) follows if

$$MQK_n^{\nu} < [\alpha(1 - q) - 1]A \exp(\alpha(1 - q)t_n) \tag{1.24a}$$

Thus $Q = Q_n$ is subject to

$$Q < [\alpha(1 - q) - 1]A^{1-\nu}M^{-1} \exp(\alpha(1 - \nu)(1 - q)t_n) \tag{1.25}$$

On the other hand, relation (1.21) follows if

$$M^{1/\mu}(\bar{q}p_0)^{-1/\mu}A^{\sigma/\mu} \exp((1 + \sigma\alpha)(1 - q)t_n/\mu) < Q \tag{1.26}$$

Hence both relations (1.14), (1.15) are satisfied if (1.18)-(1.20) hold. This completes the proof of the lemma. ■

We shall now construct an iterative method of contractor directions (Altman, 1977; 1980) as follows. Suppose that $x_0, x_1, \ldots, x_n \in V_0$ and $t_0 = 0, t_1, \ldots, t_n$ are known and satisfy the induction assumptions

$$\|x_n\|_Z < A \exp(\alpha(1 - q)t_n) = K_n \tag{1.27}$$

$$\|Px_n\| \le \|Px_0\| \exp(-(1 - q)t_n) \tag{1.28}$$

where α and A are subject to (1.19) and (1.20). Next let z_n be a solution of the equation (1.5) with $x = x_n$, $y = y_n$; find $Q = Q_n$ from Lemma 1.2 and put $z_n = x_n + h_n$ so that h_n becomes a solution of the equation (1.10) with $x = x_n$ and $y = y_n$. Now with $0 < \varepsilon_n \le 1$ to be determined put

$$x_{n+1} = x_n + \varepsilon_n h_n \qquad \text{and} \qquad t_{n+1} = t_n + \varepsilon_n,\ t_0 = 0 \tag{1.29}$$

that is,

$$x_{n+1} = (1 - \varepsilon_n)x_n + \varepsilon_n z_n$$

which justifies the term "convex approximate linearization." It follows from Lemma 1.2 that relation (1.27) holds for x_{n+1} and $t_{n+1} = t_n + \varepsilon_n$. Also relation (1.28) holds for x_{n+1} and $t_{n+1} = t_n + \varepsilon_n$. This results from (1.14) with $x = x_n$, $h = h_n$, and $\varepsilon = \varepsilon_n$, in which the right-hand term $\|Px_n\|$ is replaced by its estimate. In fact, we have

$$\begin{aligned}\|Px_{n+1}\| &\le (1 - (1 - q)\varepsilon_n)\|Px_n\| \\ &\le \exp(-(1 - q)\varepsilon_n)\|Px_n\| \\ &\le \|Px_0\| \exp(-(1 - q)(t_n + \varepsilon_n))\end{aligned}$$

by induction.

The following estimates are valid for $\{h_n\}$.

$$\|h_n\|_X \le 2C_0 p_0 \exp(-(1 - q)t_n) \tag{1.30}$$

$$\|h_n\|_Z \le \alpha(1 - q)A \exp(\alpha(1 - q)t_n) \tag{1.31}$$

In fact, relation (1.30) results from (1.15) with $x = x_n$, $h = h_n$, and (1.28). Relation (1.31) follows from (1.11) with $x = x_n$, $h = h_n$, and (1.27) and (1.24a).

Now we get from (A_0)

$$\|h_n\|_Y \le C\|h_n\|_X^{1-\bar{s}}\|h_n\|_Z^{\bar{s}}$$

and by (1.30) and (1.31) we obtain

$$\|h_n\|_Y \le CN \exp(-\lambda(1 - q)t_n) \tag{1.32}$$

where $N = (2Cp_0)^{1-\bar{s}}[\alpha(1 - q)A]^{\bar{s}}$ provided

$$\lambda = 1 - \bar{s}(1 + \alpha) > 0 \quad \text{or} \quad \alpha < (1 - \bar{s})/\bar{s} \tag{1.33}$$

1.3 LEMMA The following estimate results from (1.32).

$$\sum_{n=0}^{\infty} \varepsilon_n \|h_n\|_Y \le CN[\lambda(1 - q)]^{-1} \exp(\lambda(1 - q)) \tag{1.34}$$

Proof: We have

$$\sum_{n=0}^{\infty} \varepsilon_n \exp(-\lambda(1 - q)t_n)$$

$$= \sum_{n=0}^{\infty} (t_{n+1} - t_n) \exp(-\lambda(1 - q)t_n)$$

$$= \sum_{n=0}^{\infty} (t_{n+1} - t_n) \exp(-\lambda(1 - q)t_{n+1}) \exp(\lambda(1 - q)\varepsilon_n)$$

$$\le \exp(\lambda(1 - q)) \sum_{n=0}^{\infty} \int_{t_n}^{t_{n+1}} \exp(-\lambda(1 - q)t)\, dt$$

$$= \exp(\lambda(1 - q)) \int_0^{\infty} \exp(-\lambda(1 - q)t)\, dt$$

$$= [\lambda(1 - q)]^{-1} \exp(\lambda(1 - q))$$

In order to determine (ε_n), let c be such that $\bar{q}/q < c < 1$, and put

$$\Phi(\varepsilon,h,x) = \varepsilon^{-1}\|P(x + \varepsilon h) - (1 - \varepsilon)Px\|$$

If $\Phi(1,h_n,x_n) \le q\|Px_0\| \exp(-(1 - q)t_n)$, then put $\varepsilon_n = 1$ in (1.29). If this is not the case, then there exists $0 < \varepsilon < 1$ such that

$$\begin{aligned} cq\|Px_0\| \exp(-(1 - q)t_n) &\le \Phi(\varepsilon,h_n,x_n) \\ &\le q\|Px_0\| \exp(-(1 - q)t_n) \end{aligned} \tag{1.35}$$

and put $\varepsilon_n = \varepsilon$ in (1.29). ∎

We are now in a position to prove the following.

1.1 THEOREM In addition to the assumptions (A_0) to (A_3), suppose that conditions (1.13) and (1.18) to (1.20) are satisfied and $p_0 = \|Px_0\|$ is so small that

$$CN[\lambda(1 - q)]^{-1} \exp(\lambda(1 - q)) < r \tag{1.36}$$

with N, λ as in (1.32) and (1.33), respectively. Then the equation (1.1) has a solution x such that

$$\|x_n - x\|_Y \to 0 \qquad \text{as } n \to \infty \tag{1.37}$$

where $\{x_n\} \subset V_0$ is determined by (1.29) and $\|x - x_0\|_Y < r$.

Proof: Relations (1.29) and (1.34) imply that $\{x_n\}$ is a Cauchy sequence in Y which has a limit x, i.e., (1.37) holds. Since $t_0 = 0$ and $t_n = \Sigma_{i=0}^{n-1} \varepsilon_i$, we consider two cases: (a) $t_n \to \infty$ as $n \to \infty$. Then condition (1.28) implies

$$\|Px_n\| \to 0 \qquad \text{as } n \to \infty \tag{1.38}$$

(b) Let $\lim_{n\to\infty} t_n < \infty$. Then $\varepsilon_n \to 0$ as $n \to \infty$. But it results from (1.14a) with $x = x_n$, $h = h_n$, $\varepsilon = \varepsilon_n$, and (1.28) that

$$\Phi(\varepsilon_n, h_n, x_n) \le \varepsilon_n^{-1}\|F(x_n + \varepsilon_n h_n) - Fx_n - \varepsilon_n F'(x_n)h_n\| + \bar{q}\|Px_0\| \exp(-(1 - q)t_n) \tag{1.39}$$

Hence, we get from the first inequality (1.35) that

$$(cq - \bar{q})\|Px_0\| \exp(-(1 - q)t_n) \le \varepsilon_n^{-1}\|F(x_n + \varepsilon_n h_n) - Fx_n - \varepsilon_n F'(x_n)h_n\| \tag{1.40}$$

But the right-hand term of the inequality (1.40) converges to 0 by virtue of (A_1), since $\{h_n\}$ is bounded in Y by (1.32). Hence, relation (1.38) follows from (1.28), since $\exp(-(1 - q)t_n) \to 0$ as $n \to \infty$, by (1.40). This completes the proof of the theorem. ■

1.1 REMARK The Moser degree of approximate linearization can be eliminated by introducing the following concept.

1.2 DEFINITION Let $0 < \nu \leq 1$ and $\sigma > 0$ be given numbers. Then the linearized equation (1.2) admits approximate solutions of *direct* order or *bi-index* (ν,σ) if there exists a constant $M > 0$ with the following property. For every $x \in V_0$ and $K > 1$, if $\|x\|_Z < K$ then there exist a residual (error) vector y and a vector z such that

$$\|z\|_Z < MK^{\nu} \tag{1.41}$$

$$\|y\| < MK^{-\sigma} \tag{1.42}$$

and the equation (1.5) is satisfied. If $\nu = 1$, then σ is called the *new degree* of the approximate linearization.

The results and the proofs of this section remain valid if Definition 1.1 is replaced by Definition 1.2, provided that in Lemma 1.2 the condition (1.18) and the second inequality in (1.19) are replaced by $\alpha > \sigma^{-1}$ and (1.20) by

$$M[\alpha(1 - q) - 1]^{-1} < A^{1-\nu} \quad \text{and} \quad M(\bar{q}p_0)^{-1} < A^{\sigma} \tag{1.43}$$

2. AN ITERATIVE METHOD WITH STEPSIZE 1, GLIM-II

Let $Z \subset Y \subset X$ be Banach spaces with norms $\|\cdot\|_Z \geq \|\cdot\|_Y \geq \|\cdot\|_X$. We assume that there exist positive constants C, $\bar{s}$ with $0 < \bar{s} < 1$ such that

$$(A_0) \quad \|x\|_Y \leq C\|x\|_Y^{1-\bar{s}}\|x\|_Z^{\bar{s}}$$

Let W_0 be an open ball in Y with center $x_0 \in Z$ and radius $r_0 > 0$. Put $V_0 = W_0 \cap Z$ and let V_1 be the $\|\cdot\|_Y$-closure of V_0.

Let $P \equiv \dot{A} + F + f$ be a mapping $P\colon\ V_1 \to E$, another Banach space with norm $\|\cdot\|$ or $E = X$, where $\dot{A}$ is a nonbounded linear operator, F being strictly nonlinear, and f is Lipschitzian. If the domain of P is D, then we put $V_0 = D \cap W_0 \cap Z$. Consider the nonlinear operator equation

$$Px \equiv \dot{A}x + Fx + fx = 0 \tag{2.1}$$

We assume that F is differentiable, i.e., for each $x \in V_0$, there exists a linear operator $F'(x)$ such that

$$\varepsilon^{-1}\|F(x + \varepsilon h) - Fx - \varepsilon F'(x)h\| \to 0$$

as $\varepsilon \to 0+$, where $h \in Y$.

2.1 DEFINITION (Notation) Let $\mu > 0$, $0 \leq \nu < 1$, $\sigma \geq 0$ be given. Then the linearized equation

$$\dot{A}z + F'(x)z + Fx - F'(x)x + f(x) = 0 \tag{2.2}$$

admits approximate solutions of order (μ,ν,σ) if there exists a constant $M > 0$ with the following property. For every $x \in V_0$, $K > 1$ and $Q > 1$, if $\|x\|_Z < K$ then there exist a residual (error) vector y and a vector z such that

$$\|z\|_Z \leq MQK^{\nu} \tag{2.3}$$

$$\|y\| \leq MQ^{-\mu}K^{\sigma} \tag{2.4}$$

and

$$\dot{A}z + F'(x)z + Fx - F'(x)x + fx + y = 0 \tag{2.5}$$

We make the following assumptions.

(A_1) There exists a constant $C_1 > 1$ such that

$$\|F(x + h) - Fx - F'(x)h\| \leq C_1\|h\|_X\|h\|_Y \tag{2.6}$$

for $x \in V_0$.

P is closed on V_1.

(A_2) There exists a constant $C_0 > 0$ with the following property. For $x \in V_0$ and $g \in E$ if h is a solution of the equation

$$\dot{A}h + F'(x)h + g = 0 \tag{2.7}$$

then

$$\|h\|_X \leq C_0\|g\| \tag{2.8}$$

There exists a constant $q_0 > 0$ such that

$$\|f(x + h) - fx\| \leq q_0\|h\|_X \quad \text{and} \quad q_0C_0 < 1,\ x \in V_0 \tag{2.9}$$

Now for $x \in V_0$, let z be a solution of equation (2.5) and put $z = x + h$. Then obviously h is a solution of the equation

$$\dot{A}h + F'(x)h + Px + y = 0 \tag{2.10}$$

and we get from (2.3), (2.4), and (2.7) with g replaced by $Px + y$

$$\|h\|_Z \leq \|x\|_Z + MQK^{\nu} \tag{2.11}$$

and

$$\|h\|_X \leq C_0(\|Px\| + MQ^{-\mu}K^{\sigma}) \tag{2.12}$$

2.1 LEMMA Let $x \in V_0$, and let

$$(C_1 r_0 + q_0)C_0 < 1 \tag{2.13}$$

Let h be a solution of equation (2.10). If $K > 1$, $\|x\|_Z < K$, and $Q > 1$ are such that

$$\|y\| \leq MQ^{-\mu}K^{\sigma} \leq \bar{q}\|Px\| \tag{2.14}$$

then

$$\|P(x + h)\| \leq q\|Px\| \tag{2.15}$$

$$\|h\|_X \leq C_0(1 + \bar{q})\|Px\| \tag{2.16}$$

provided that

$$(C_1 r_0 + q_0)C_0(1 + 2\bar{q}) < q < 1 \qquad \bar{q} < q \tag{2.17}$$

Proof: We use the following identity

$$\begin{aligned} P(x + h) = {} & [F(x + h) - Fx - F'(x)h] + [f(x + h) - fx] \\ & + [\dot{A}h + F'(x)h + Px + y] - y \end{aligned}$$

Hence, we obtain from (2.6), (2.12), (2.14), and (2.10),

$$\begin{aligned} \|P(x + h)\| &\leq \|F(x + h) - Fx - F'(x)h\| + q_0\|h\|_X + \|y\| \\ &\leq (C_1 r_0 + q_0)\|h\|_X + \bar{q}\|Px\| \\ &\leq (C_1 r_0 + q_0)C_0(1 + 2\bar{q})\|Px\| \end{aligned}$$

and relation (2.15) follows. Relation (2.16) results from (2.8) and (2.14). ■

2.2 LEMMA For $x_n \in V_0$, let

$$\|x_n\|_Z < Aq^{-\alpha n} = K_n \tag{2.18}$$

$$\|Px_n\| \le p_0 q^n \tag{2.19}$$

where $p_0 = \|Px_0\|$, and α, A are chosen so as to satisfy the relations

$$\alpha[\mu(1 - \nu) - \sigma] > 1 \tag{2.20}$$

$$M^{1+1/\mu}(\bar{q}p_0)^{-1/\mu}(q^{-\alpha} - 2)^{-1} < A^{1-\nu-\sigma/\mu} \qquad q < 2^{-1/\alpha} \tag{2.21}$$

Then there exists a number $Q = Q_n$ such that

$$MQ_n^{-\mu}K_n^{\sigma} \le \bar{q}p_0 q^n \tag{2.22}$$

and

$$\|x_n + h_n\|_Z < Aq^{-\alpha(n+1)} = K_{n+1} \tag{2.23}$$

where h_n is a solution of equation (2.10) with $x = x_n$, i.e.,

$$\dot{A}h_n + F'(x_n)h_n + Px_n + y_n = 0 \tag{2.24}$$

which satisfies relations (2.11) and (2.12).

Proof: Relation (2.12) with $x = x_n$ implies

$$\|x_n + h_n\|_Z \le \|x_n\|_Z + \|h_n\|_Z \le 2\|x_n\|_Z + MQ_nK_n^{\nu}$$

since $h_n = z_n - x_n$. Hence (2.23) follows if

$$2Aq^{-\alpha n} + MQ_nK_n^{\nu} < Aq^{-\alpha(n+1)} \tag{2.25}$$

or

$$Q_n < M^{-1}A^{1-\nu}q^{-\alpha(1-\nu)n}(q^{-\alpha} - 1) \tag{2.26}$$

On the other hand, (2.22) holds if

$$M^{1/\mu}(\bar{q}p_0)^{-1/\mu}A^{\sigma/\mu}q^{-(\alpha\sigma+1)n/\mu} < Q_n \tag{2.27}$$

Hence, relations (2.26) and (2.27) are satisfied if conditions (2.20) and (2.21) hold.

We shall now construct the following iterative method. With x_0 as above, assume that $x_0, x_1, \ldots, x_n$ are known and satisfy the relations (2.18) and (2.19) of Lemma 2.2. Then put

$$x_{n+1} = z_n = x_n + h_n \tag{2.28}$$

where z_n and h_n are solutions of the equations (2.5) with $x = x_n$, $y = y_n$, and (2.10), respectively, i.e., z_n is an approximate solution of the linearized equation (2.2). It results from (2.23) that relation (2.18) holds true for $n + 1$, and relation (2.15) with $x = x_n$, $h = h_n$ implies that condition (2.19) is still satisfied for $n + 1$. ■

2.3 LEMMA The following estimate holds

$$\sum_{n=0}^{\infty} \|h_n\|_Y \le N/(1 - q^{\beta}) \tag{2.29}$$

where $N = C[C_0(1 + \bar{q})p_0]^{1-\bar{s}}[A(q^{-\alpha} - 1)]^{\bar{s}}$, $\beta = [1 - \bar{s}(1 + \alpha)] > 0$.

Proof: We have by (2.16) with $h = h_n$, $x = x_n$, and (2.19),

$$\|h_n\|_X \le C_0(1 + \bar{q})p_0q^n \tag{2.30}$$

with $\bar{q}$ from (2.17). It results from (2.11) with $x = x_n$, $h = h_n$, (2.18), and (2.25) that

$$\begin{aligned}\|h_n\|_Z &\le \|x_n\|_Z + MQ_nK_n^{\nu} \\ &\le Aq^{-n} + Aq^{-\alpha(n+1)} - 2Aq^{-n} \\ &= A(q^{-\alpha} - 1)q^{-\alpha n}\end{aligned}$$

Hence, with $q < 2^{-1/\alpha}$ we get

$$\|h_n\|_Z \le A(q^{-\alpha} - 1)q^{-\alpha n} \tag{2.31}$$

Relations (2.30), (2.31), and (A_0) imply

$$\|h_n\|_Y \le C\|h_n\|_X^{1-\bar{s}}\|h_n\|_Z^{\bar{s}} \le Nq^{\beta} \tag{2.32}$$

where α is chosen so that

$$\beta = [1 - \bar{s}(1 + \alpha)] > 0 \qquad \alpha < (1 - \bar{s})/\bar{s},\ q < 2^{-1/\alpha} \tag{2.33}$$

and in addition, conditions (2.20) and (2.21) are satisfied. ■

We are now in a position to prove the following.

2.1 THEOREM In addition to assumptions $(A_0\text{-}A_3)$, suppose that

$$[\mu(1 - \nu) - \sigma]^{-1} < \frac{1 - \bar{s}}{\bar{s}} \tag{2.34}$$

and α, A are chosen so as to satisfy conditions (2.33) and (2.21), where $\bar{q}$ and q are subject to (2.17). If

$$N(1 - q^{\beta}) < r_0 \tag{2.35}$$

then equation (2.1) has a solution x, such that

$$\|x_n - x\|_Y \to 0 \qquad \text{as } n \to \infty \tag{2.36}$$

and $\{x_n\} \subset V_0$.

Proof: The convergence of $\{x_n\}$ results from (2.28) and (2.29). Conditions (2.29) and (2.35) also imply $\{x_n\} \subset V_0$. Since $\|Px_n\| \to 0$ as $n \to \infty$ by (2.19), and P is closed on V_1, it follows that $Px = 0$. This completes the proof of the theorem. ■

2.1 REMARK Remark 1.1 in Section 1 also applies to this section.

3. A RAPIDLY CONVERGENT ITERATION METHOD, GLIM-III

Let $Z \subset Y \subset X$ be Banach spaces with norms $\|\cdot\|_Z \ge \|\cdot\|_Y \ge \|\cdot\|_X$. We assume that there exist positive constants $C, \bar{s}$ with $0 < \bar{s} < 1$ such that

$$(A_0) \quad \|x\|_Y \le C\|x\|_X^{1-\bar{s}}\|x\|_Z^{\bar{s}}$$

Let W_0 be an open ball in Y with center $x_0 \in Z$ and radius $r_0 > 0$. Put $V_0 = W_0 \cap Z$ and let V_1 be the $\|\cdot\|_Y$-closure of V_0.

Let $P \equiv A + F$ be a mapping $P: V_1 \to E$, another Banach space with norm $\|\cdot\|$ or $E = X$, where A is a nonbounded linear operator, F being strictly nonlinear. If the domain of P is D, then we put $V_0 = D \cap W_0 \cap Z$. Consider the nonlinear operator equation

$$Px \equiv Ax + Fx = 0 \tag{3.1}$$

We assume that F is differentiable, i.e., for each $x \in V_0$, there exists a linear operator $F'(x)$ such that

$$\varepsilon^{-1}\|F(x + \varepsilon h) - Fx - \varepsilon F'(x)h\| \to 0$$

as $\varepsilon \to 0+$, where $h \in Y$.

3.1 DEFINITION (Notation) Let $\mu > 0$, $\nu \geq 0$, and $\sigma \geq 0$ be given constants. Then the linearized equation

$$Az + F'(x)z + Fx - F'(x)x = 0 \tag{3.2}$$

admits approximate solutions of order (μ,ν,σ) if there exists a constant $M > 0$ with the following property. For every $x \in V_0$, $K > 1$, $Q > 1$, if $\|x\|_Z < K$ then there exist a residual (error) vector y and a vector z such that

$$\|z\|_Z \leq MQK^{\nu} \tag{3.3}$$

$$\|y\| \leq MQ^{-\mu}K^{\sigma} \tag{3.4}$$

and

$$Az + F'(x)z + Fx - F'(x)x + y = 0 \tag{3.5}$$

We make the following assumptions.

(A_1) P is closed on V_1 and F' satisfies the following condition

$$\|F(x + h) - Fx - F'(x)h\| \leq \bar{M}\|h\|_X^{2-\beta}\|h\|_Z^{\beta} \tag{3.6}$$

for some $\bar{M} > 0$, $0 \leq \beta < 1$, and all $x \in V_0$.

(A_2) There exists a constant C_0 with the following property. If h is a solution of the equation

$$Ah + F'(x)h + g = 0 \tag{3.7}$$

then

$$\|h\|_X \le C_0\|g\| \tag{3.8}$$

(A_3) The linearized equations (3.2) admit approximate solutions in the sense of Definition 3.1 with $\nu = \sigma = 1 + \alpha$, $0 < \alpha < 1$, $\mu > 1$.

(A_4) Let $\alpha, \tau, \lambda < \mu$ be such that

$$1 < (\mu - \lambda)^{-1}(1 + \alpha(1 + \lambda) + \mu) < \tau < 2 - \alpha_0 < 2 - \alpha \tag{3.9}$$

$$0 < 2\lambda < [\mu(1 - \alpha_0) - (1 + \alpha_0)]/(1 + \alpha_0) \tag{3.10}$$

where $0 < \alpha < \alpha_0 < 1$ is such that

$$\mu > (1 + \alpha_0)/(1 - \alpha_0) > 1 \tag{3.11}$$

and let β be such that

$$0 < \beta < \mu\lambda(\alpha_0 - \alpha)[(1 + \alpha)(1 + \mu) + \lambda(2 + \mu)]^{-1} < 1 \tag{3.12}$$

3.1 REMARK One can put $\alpha_0 = (\mu - 1)/(\mu + 3)$ in (3.9)-(3.12). Then instead of (3.10) we have

$$0 < 2\lambda < (\mu - 1)/(\mu + 1) \tag{3.10a}$$

The following method generalizes some of Moser's (1966) essential technique, which he developed for solving nonlinear operator equations in a scale of Banach spaces, especially in case of "loss of derivatives." Since we assume $\nu = \sigma = 1 + \alpha$ $(0 < \alpha < 1)$ instead of $\nu = \sigma = 1$ in Moser's case, the scope of applications of the following method is enhanced (see Chapter 2). The method has been applied to nonlinear evolution equations in Banach spaces (Altman, 1985b).

Let $K_{n+1} = K_n^{\tau+\alpha}$ $(1 < \tau < 2)$ and put

$$x_{n+1} = z_n = x_n + h_n \qquad x_n \in V_0 \qquad n = 0, 1, \ldots \tag{3.13}$$

where z_n is a solution of equation (3.5) with $x = x_n$, $y = y_n$. Hence, h_n is obviously a solution of the equation

$$Ah + F'(x)h + Px + y = 0 \tag{3.14}$$

with $x = x_n$, $y = y_n$. It will be shown that Q_n from (3.3) and (3.4) can be chosen so that the following induction assumptions are satisfied

$$\|x_n\|_Z < K_n \qquad x_n \in V_0 \tag{3.15}$$

$$\|Px_n\| \leq K_n^{-\lambda} \tag{3.16}$$

for some $\lambda > 0$ to be determined.

3.1 THEOREM Suppose that assumptions (A_0) through (A_4) are satisfied. Then there exists a constant $K_0(M,\beta,\mu,\lambda,\alpha) > 1$ such that if

$$\|Px_0\| < K_0^{-\lambda} \qquad \|x_0\|_Z < K_0 \qquad \text{and} \qquad C(2C_0)^{1-\bar{s}} \sum_{n=0}^{\infty} K_n^{-\delta} < r_0 \tag{3.17}$$

where $\delta = (1 - \bar{s})\lambda - \bar{s}\tau > 0$ provided

$$\bar{s} < \lambda/(\lambda + 2) \tag{3.18}$$

and $K_{n+1} = K_n^{\tau+\alpha}$ with τ,α as in (3.9), then the equation (3.1) has a solution x which is a limit of $\{x_n\}$, and

$$\|x_n - x\|_Y \to 0 \qquad \text{as } n \to \infty \tag{3.19}$$

where $\{x_n\}$ is determined by (3.13).

Proof: To verify the assumption (3.15) we use (3.13) and (3.3) with $z = z_n$, $K = K_n$, $\nu = 1 + \alpha$, and choose $Q = Q_n$ so as to satisfy

$$2MQK_n^{1+\alpha} < K_{n+1} = K_n^{\tau+\alpha} \tag{3.20}$$

Then obviously

$$\|x_{n+1}\|_Z = \|z_n\|_Z < MQK_n^{1+\alpha} < K_{n+1} \tag{3.21}$$

To verify the induction assumption (3.16) we use relation (3.6) with $x = x_n$, $h = h_n$, where h_n satisfies (3.14) with $x = x_n$, $y = y_n$, and the following identity

$$\begin{aligned}
Px_{n+1} &= P(x_n + h_n) \\
&= [F(x_n + h_n) - Fx_n - F'(x_n)h_n] \\
&= [Ah_n + F'(x_n)h_n + Px_n + y_n] - y_n
\end{aligned}$$

Hence, we get by virtue of (3.22), (3.24), and (3.4),

$$\begin{aligned}
\|Px_{n+1}\| &\le \bar{M}\|h_n\|_X^{2-\beta}\|h_n\|_Z^{\beta} + \|y_n\| \\
&\le M(2C_0)^{2-\beta}K_n^{-\lambda(2-\beta)}(2M)^{\beta}(QK_n^{1+\alpha})^{\beta} + MQ^{-\mu}K_n^{1+\alpha} \\
&\le c[K_n^{-\lambda(2-\beta)}(QK_n^{1+\alpha})^{\beta} + Q^{-\mu}K_n^{1+\alpha}]
\end{aligned}$$

for some constant $c > M$, since

$$\begin{aligned}
\|h_n\|_X &\le C_0(\|Px_n\| + \|y_n\|) \\
&\le C_0(K_n^{-\lambda} + MQ^{-\mu}K_n^{1+\alpha})
\end{aligned}$$

Hence,

$$\|h_n\|_X \le 2C_0K_n^{-\lambda} \tag{3.22}$$

provided that Q is chosen so as to satisfy

$$MQ^{-\mu}K_n^{1+\alpha} < K_n^{-\lambda} \tag{3.23}$$

Since $z_n = x_n + h_n$, we get

$$\begin{aligned}
\|h_n\|_Z &\le \|x_n\|_Z + \|z_n\|_Z \\
&\le K_n + MQK_n^{1+\alpha}
\end{aligned}$$

Hence,

$$\|h_n\|_Z \le 2MQK_n^{1+\alpha} < K_{n+1} \tag{3.24}$$

by virtue of (3.20). Thus (3.16) can be verified if Q can be chosen so as to satisfy

$$c[K_n^{-\lambda(2-\beta)}(QK_n^{1+\alpha})^{\beta} + Q^{-\mu}K_n^{1+\alpha}] < K_{n+1}^{-\lambda} \tag{3.25}$$

Therefore, assuming that K_0 is sufficiently large, and by virtue of (3.20), (3.23), and (3.25), it suffices to show that the following inequalities can be satisfied for some $Q > 1$.

$$cK^{1+\alpha}Q < K_{n+1} \tag{3.26}$$

$$c(K^{1+\alpha}Q)^{\beta}K_n^{-\lambda(2-\beta)} < K_{n+1}^{-\lambda} \tag{3.27}$$

$$cK_n^{1+\alpha}Q^{-\mu} < K_{n+1}^{-\lambda} \tag{3.28}$$

Using $K_{n+1} = K_n^{\tau+\alpha}$ and comparing the exponents in (3.26) and (3.28) yields

$$1 + \alpha + \lambda(\tau + \alpha) < \mu(\tau - 1) \tag{3.29}$$

and a comparison of the exponents in (3.27) and (3.28) gives

$$\mu^{-1}[1 + \alpha + \lambda(\tau + \alpha)] < \beta^{-1}[\lambda(2 - \beta) - \lambda(\tau + \alpha) - (1 + \alpha)\beta] \tag{3.30}$$

Inequality (3.30) can be written as

$$\mu^{-1}[(1 + \alpha)(1 + \mu) + \lambda(\tau + \alpha + \mu)] < \beta^{-1}[2\lambda - \lambda(\tau + \alpha)]$$

or

$$\beta < \mu\lambda[2 - (\tau + \alpha)]/[(1 + \alpha)(1 + \mu) + \lambda(\tau + \alpha + \mu)] \tag{3.31}$$

But (3.12) and (3.9) imply (3.31), since $\alpha_0 - \alpha < 2 - (\tau + \alpha)$, by (3.9). On the other hand, inequality (3.29) can be replaced by the stronger one

$$\frac{1 + \alpha_0(1 + \lambda) + \mu}{\mu - \lambda} < \tau \tag{3.32}$$

Now if $\lambda > 0$ can be chosen so as to satisfy

$$\frac{1 + \alpha_0(1 + \lambda) + \mu}{\mu - \lambda} < 2 - \alpha_0 \tag{3.33}$$

then one can find $1 < \tau < 2 - \alpha_0$ such that relation (3.9) is satisfied. But (3.33) can be written as (3.10), and $\lambda > 0$ if

$\mu > (1 + \alpha_0)/(1 - \alpha_0)$. To determine α_0 let $1 < k < \mu$ and put $k = (1 + \alpha_0)/(1 - \alpha_0)$, i.e., $\alpha_0 = (k - 1)/(k + 1)$. Then α is subject to $0 < \alpha < \alpha_0 < 1$, and (3.11) holds. A simple choice is $k = (1 + \mu)/2$. Then $\alpha_0 = (\mu - 1)/(\mu + 3) > \alpha$, and instead of (3.10) we get (3.10a). Thus $Q > 1$ can be chosen so as to satisfy (3.26) to (3.28) if K_0 is large enough.

Finally, we get from (A_0), (3.22), and (3.24),

$$\begin{aligned}\|h_n\|_X &\le C\|h_n\|_X^{1-\bar{s}}\|h_n\|_Z^{\bar{s}} \\ &\le C(2C_0K_n^{-\lambda})^{1-\bar{s}}K_n^{\tau\bar{s}} = C(2C_0)^{1-\bar{s}}K_n^{-\delta}\end{aligned}$$

where $\delta > 0$ is from (3.17). Hence,

$$\sum_{n=0}^{\infty} \|h_n\|_Y \le C(2C_0)^{1-\bar{s}} \sum_{n=0}^{\infty} K_n^{-\delta} < r_0 \tag{3.34}$$

It follows from (3.34) that $\|x_n\|_Y < r_0$, i.e., $x_n \in V_0$, and (3.19) holds for some $x \in V_1$ which is a solution of the equation (3.1), by virtue of (3.16) and (A_1). It is clear that relation (3.34) is satisfied if K_0 is sufficiently large. ∎

The above theorem does not indicate how to construct approximate solutions of the linearized equations in the sense of Definition 3.1. One way of obtaining such approximations is to make use of the concept of smoothing operators introduced by Nash (1956) and developed by Moser (1966).

3.2 REMARK Assumption (3.6) is essential in our discussion. Therefore GLIM-III is not applicable to the following more general equation

$$Ax + Fx + fx = 0$$

where fx is only Lipschitzian and (3.6) holds for F so that (3.6) is not satisfied for $F + f$. In this case GLIM-II is applicable, and moreover, GLIM-I also applies even if f is locally Lipschitzian.

Consider now the case where $\nu = \sigma = 1$ in Definition 3.1. Here instead of (A_4) we make the following assumption.

(A'_4) Suppose that

$0 < \lambda + 1 < (\mu + 1)/2$

and

$$0 < \beta < \frac{\lambda}{\lambda + 1} \frac{\mu}{\mu + 1} \left(1 - 2 \frac{\lambda + 1}{\mu + 1}\right) \tag{3.35}$$

where β is from (1.6). Then τ is a number satisfying the following

$$1 < \left(1 - \frac{\lambda + 1}{\mu + 1}\right)^{-1} < \tau < 2 \tag{3.36}$$

3.2 THEOREM Theorem 1.1 remains valid if (A_4) is replaced by (A'_4), $\nu = \sigma = 1$, and $K^{\tau}_{n+1} = K_n$, $n = 0, 1, \ldots$.

Proof: The proof is exactly the same as that of Theorem 3.1 provided that the inequalities (3.26) to (3.28) are replaced by the following ones.

$$cK_nQ < K_{n+1} = K_n^{\tau} \tag{3.37}$$

$$c(K_nQ)^{\beta}K_n^{-\lambda(2-\beta)} < K_{n+1}^{-\lambda} \tag{3.38}$$

$$cK_nQ^{-\mu} < K_{n+1}^{-\lambda} \tag{3.39}$$

Moser (1966) has shown that $Q > 1$ satisfying (3.37) to (3.39) can be found. In fact, by comparing the exponents in (3.37) and (3.39) one obtains

(a) $\tau\lambda + 1 < \mu(\tau - 1)$

A comparison of the exponents in (3.38) and (3.39) gives

(b) $\tau\lambda + 1 < \mu\left(-1 + \lambda \dfrac{2 - \beta}{\beta} - \tau \dfrac{\lambda}{\beta}\right)$

which can be written as

(c) $\tau\lambda + 1 < \mu\left[-(1 + \lambda) + (2 - \tau) \dfrac{\lambda}{\beta}\right]$

which can be replaced by a stronger inequality

(d) $\tau(\lambda + 1) < \mu\left[-\tau(\lambda + 1) + (2 - \tau) \dfrac{\lambda}{\beta}\right]$

But (d) is equivalent to

$$\frac{\tau}{\tau - 1} > \frac{\lambda + 1}{\lambda} \cdot \frac{\mu + 1}{\mu} \beta$$

or

$$\frac{\lambda + 1}{\lambda} \cdot \frac{\mu + 1}{\mu} \beta < \frac{2}{\tau} - 1 < 2\left(1 - \frac{\lambda + 1}{\mu + 1}\right) - 1 = 1 - 2\frac{\lambda + 1}{\mu + 1}$$

since $1/\tau < (\mu - \lambda)/(\mu + 1)$, by (a). This shows that a number τ can be found that satisfies (3.36) if (3.34) and (3.35) hold. Hence, (3.37) to (3.39) are compatible if K_0 is large enough, where K_0 depends on M, β, μ, λ. ∎

3.3 REMARK Definition 1.2 (see Remark 1.1 in Section 1) can also be applied to this section. Then instead of the system (3.37) to (3.39) we obtain the following:

$$K_n < K_{n+1} = K_n^{\tau} \tag{3.40}$$

$$K_n^{\beta} K_n^{-\lambda(2-\beta)} < K_{n+1}^{-\lambda} = K_n^{-\tau\lambda} \tag{3.41}$$

$$K_n^{-\sigma} < K_{n+1}^{-\lambda} \tag{3.42}$$

By comparing the exponents in (3.41) and (3.42), one obtains

$$\frac{\beta}{2 - \tau - \beta} < \lambda < \frac{\sigma}{\tau} \tag{3.43}$$

which yields

$$\tau < \frac{(2 - \beta)\sigma}{\beta + \sigma} \tag{3.44}$$

The inequality (3.44) and

$$\tau > 1 \tag{3.45}$$

yield

$$\beta < \frac{\sigma}{1 + \sigma} \tag{3.46}$$

Thus, the inequality (3.46) and

$$1 < \tau < \frac{(2 - \beta)\sigma}{\sigma + \beta} \tag{3.47}$$

are sufficient for (3.40) to (3.42) to be satisfied if (3.45) holds or if

$$\frac{\beta}{2 - \tau - \beta} < \lambda < \frac{\sigma + \beta}{2 - \beta} < \frac{\sigma}{\tau}$$

2
Smoothing Operators Combined With Elliptic Regularization and the Degree of Elliptic Regularization

This chapter shows how to construct approximate solutions of the linearized equations. To this end we use smoothing operators invented by Nash (1956). The best results can be obtained by a combination of smoothing operators and elliptic regularization, which is also known as artificial viscosity. In this way no "gain of derivatives" is needed for the estimate of the solution of the linearized equation. Elliptic regularization can also be used without smoothing operators. But then an appropriate gain of derivatives is needed for the estimate of the solution of the linearized equations. The degree of elliptic regularization, which is a new concept, is also introduced. Of fundamental importance is the key Lemma 1.6, which is used in all three sections. Moser's (1966) degree of approximate linearization can be replaced by a new concept of the degree. In this case Lemma 1.6a replaces Lemma 1.6.

Having constructed the approximate solutions of the linearized equations one then applies the results of Chapter 1 in order to prove their convergence to a solution of the operator equation. All three methods (GLIM) are examined. An example which shows existence of smoothing operators in Sobolev-Hilbert spaces is given in Section 1, Chapter 4.

1. GLIM-I IN A SCALE OF BANACH SPACES

Let $\{X_j\}$ with $0 \le j \le p$ be a scale of Banach spaces with increasing norms such that $i < j$ implies $X_j \subset X_i$ and $\|x\|_j \ge \|x\|_i$, and let $0 < m_1 < m_2 < \bar{p} < p$. We make the following assumptions.

(A_0) We assume that there exists a one-parameter family of linear smoothing operators S_θ, $\theta \ge 1$, such that

$$\|(I - S_\theta)x\|_0 \le C\theta^{-m_1}\|x\|_{m_1} \qquad m_1 \ge 0$$

$$\|(I - S_\theta)x\|_{m_1} \le C\theta^{-(m_2-m_1)}\|x\|_{m_2}$$

$$\|S_\theta x\|_p \le C\theta^{(p-m_2)}\|x\|_{m_2}$$

for some constant $C > 0$, where I is the identity mapping. We also assume

$$LS_\theta = S_\theta L \qquad \text{for all } \theta \ge 1$$

and

$$\|x\|_j \le C\|x\|_i^{1-\lambda}\|x\|_p^\lambda \tag{1.0}$$

for $j = (1 - \lambda)i + \lambda p$; $0 \le \lambda \le 1$.

Let $W_0 \subset X_s$ be an open ball with center $x_0 \in X_p$ and radius $r_0 > 0$. Put $V_0 = W_0 \cap X_p$ and let V_s be the $\|\cdot\|_s$-closure of V_0 with s to be determined later.

Consider the nonlinear operator equation

$$Px \equiv Lx + Fx + fx = 0 \tag{1.1}$$

where P is closed on V_s and assumes its values in the same scale of Banach spaces provided that $L \neq 0$. If the domain of P is $\mathcal{D} \subset X_s$, then we put $V_0 = \mathcal{D} \cap W_0 \cap X_p$.

We make the following assumptions.

(A_1) F is differentiable; i.e., for each $x \in V_0$, there exists a linear operator $F'(x)$ such that

$$\varepsilon^{-1}\|F(x + \varepsilon h) - Fx - \varepsilon F'(x)h\|_0 \to 0 \qquad \text{as } \varepsilon \to 0+ \tag{1.2}$$

There exists a constant $C > 0$ such that

$$\|F'(x)h\|_{m_1} \leq C\|h\|_{m_2} \tag{1.3}$$

$$\|F'(x)h\|_0 \leq C\|h\|_{m_1} \tag{1.4}$$

for all $x \in V_0$.

Let $\{x_n\} \subset V_0$ be a Cauchy sequence in X_s and let $\{h_n\}$ be bounded in X_s, then $\varepsilon_n \to 0+$ implies that

$$\varepsilon_n^{-1}\|F(x_n + \varepsilon_n h_n) - Fx_n - \varepsilon_n F'(x_n)h_n\|_0 \to 0 \tag{1.5}$$

There exists a constant $C > 0$ such that $x \in V_0$ with $\|x\|_{\bar{p}} \geq 1$ imply

$$\|Fx - F'(x)x + fx\|_{m_2} \leq C\|x\|_{\bar{p}} \tag{1.6}$$

and

$$\|fx\|_{m_2} \leq C\|x\|_{\bar{p}} \tag{1.7}$$

(A_2) There exists a constant q_0 such that

$$\|f(x + \varepsilon h) - fx\|_0 \leq \varepsilon q_0\|h\|_0 \tag{1.8}$$

for all $x \in V_0$.

(A_3) There exists a linear (regularizing) operator $L = L(\eta)$ such that

$$\|Lx\|_0 \leq C\|x\|_{m_2} \tag{1.9}$$

for some constant $C > 0$.

The linearized equation

$$Lz + F'(x)z + \eta Lz + Fx - F'(x)x + fx = 0 \tag{1.10}$$

with small $0 < |\eta| < 1$ has a solution $\bar{z}$ such that

$$\|\bar{z}\|_{m_2} \leq C|\eta|^{-k}\|Fx - F'(x)x + fx\|_{m_2} \tag{1.11}$$

for some $\bar{k} > 0$ to be determined and $C > 0$.

(A_4) For $x \in V_0$ if h is a solution of the equation

$$Lh + F'(x)h + g = 0 \tag{1.12}$$

then

$$\|h\|_0 \le C_0\|g\|_0 \tag{1.13}$$

for some constant $C_0 > 0$. In order to simplify the notation we assume that $0 < \eta < 1$.

For $x \in V_0$, let $\bar{z}$ be a solution of the equation (1.10) satisfying (1.11). Then the following estimates hold.

1.1 LEMMA The following relations hold for $x \in V_0$ with $\|x\|_p < K$, $K > 1$; $0 < \eta < 1$.

$$\|L(I - S_\theta)\bar{z}\|_0 \le M_1(\theta^{-m_1} + \eta)\eta^{-\bar{k}}K^{\bar{\nu}} \tag{1.14}$$

$$\|F'(x)(I - S_\theta)\bar{z}\|_0 \le M_2\theta^{-(m_2-m_1)}\eta^{-\bar{k}}K^{\bar{\nu}} \tag{1.15}$$

$$\|z\|_p \equiv \|S_\theta\bar{z}\|_p \le M_3\theta^{p-m_2}\eta^{-\bar{k}}K^{\bar{\nu}} \tag{1.16}$$

for some M_1, M_2, M_3, and $z = S_\theta\bar{z}$, $\bar{\nu} = (\bar{p} - \rho)/(\bar{p} - \rho)$ with $\rho \le s$ to be determined, $0 < \bar{k} < 1$.

Proof: We get from (A_0), (1.3), and (1.6),

$$\|L(I - S_\theta)\bar{z}\|_0 = \|(I - S_\theta)[F'(x)\bar{z} + \eta L\bar{z} + Fx - F'(x)x + fx]\|_0$$

$$\le C\theta^{-m_1}[\|F'(x)\bar{z}\|_{m_1} + \|Fx - F'(x)x + fx\|_{m_1}] + C\eta\|L\bar{z}\|_0$$

$$\le C\theta^{-m_1}C\|\bar{z}\|_{m_2} + C\eta\|\bar{z}\|_{m_2} + C\theta^{-m_1}\|x\|_{\bar{p}}$$

Hence, relation (1.14) follows, since

$$\|x\|_{\bar{p}} \le C\|x\|_\rho^{1-\bar{\nu}}\|x\|_p^{\bar{\nu}} \le Cr^{1-\bar{\nu}}K^{\bar{\nu}}$$

and

$$\|\bar{z}\|_{m_2} \le C_1\eta^{-\bar{k}}\|x\|_{\bar{p}} \le C_2\eta^{-\bar{k}}\|x\|_p^{\bar{\nu}} \tag{1.17}$$

We have

$$\|F'(x)(I - S_\theta)\bar{z}\|_0 \le C\|(I - S_\theta)\bar{z}\|_{m_1} \le C\theta^{-(m_2-m_1)}\|\bar{z}\|_{m_2}$$

and (1.15) follows from (1.17).

By virtue of (A_0), we get

$$\|S_\theta\bar{z}\|_p \le C\theta^{p-m_2}\|\bar{z}\|_{m_2}$$

and (1.16) results from (1.17). ∎

Consider the linearized equation

$$Lz + F'(x)z + Fx - F'(x)x + fx = 0 \tag{1.18}$$

and put $z = S_\theta\bar{z}$, where $\bar{z}$ is a solution of the linearized equation (1.10) satisfying relation (1.11).

1.2 LEMMA The following estimate holds

$$\|Lz + F'(x)z + Fx - F'(x)x + fx\|_0 \le M_4(\theta^{-m} + \eta)\eta^{-\bar{k}}K^{\bar{\nu}} \tag{1.19}$$

for $x \in V_0$ with $\|x\|_p < K$ and some $M_4 > 0$, where

$$m = \min(m_1, m_2 - m_1) \tag{1.20}$$

Proof: We have

$$\|Lz + F'(x)z + Fx - F'(x)x + fx\|_0$$

$$\le \|L\bar{z} + F'(x)\bar{z} + \eta L\bar{z} + Fx - F'(x)x + fx\|_0$$

$$+ \|L(I - S_\theta)\bar{z}\|_0 + \|F'(x)(I - S_\theta)\bar{z}\|_0 + \eta\|L\bar{z}\|_0$$

and (1.19) follows, by virtue of Lemma 1.1, and (1.10), (1.11), and (1.20). ∎

Now suppose there exist ν,μ with

$$\bar{\nu} < \nu < 1 \quad \text{and} \quad \mu(1 - \nu) - \nu > 0 \tag{1.21}$$

such that for $K > 1$, $Q > 1$, one can find $0 < \eta < 1$ and $\theta > 1$ to satisfy

$$\theta^{p-m_2} \eta^{-\bar{k}} K^{\bar{\nu}} < QK^{\nu} \tag{1.22}$$

and

$$(\theta^{-m} + \eta)\eta^{-\bar{k}} K^{\bar{\nu}} < Q^{-\mu} K^{\nu} \tag{1.23}$$

Then relation (1.19) can be written as

$$\|Lz + F'(x)z + Fx - F'(x)x + fx\|_0 \leq 2M_4 Q^{-\mu} K^{\nu} \tag{1.24}$$

by (1.21) to (1.23).

1.1 DEFINITION Let $\mu > 0$, $0 \leq \nu < 1$, $\sigma \geq 0$ be given numbers. Then the linearized equation (1.18) admits approximate solutions of order (μ,ν,σ) if there exists a constant $M > 0$ with the following property. For every $x \in V_0$, $K > 1$ and $Q > 1$ if $\|x\|_p < K$, then there exist a residual (error) vector y and a vector z such that

$$\|z\|_p \leq MQK^{\nu} \tag{1.25}$$

$$\|y\|_0 \leq MQ^{-\mu} K^{\sigma} \tag{1.26}$$

and

$$Lz + F'(x)z + Fx - F'(x)x + fx + y = 0 \tag{1.27}$$

1.3 LEMMA The linearized equation (1.18) admits approximate solutions of order (μ,ν,σ) in the sense of Definition 1.1 with

$$\sigma = \nu \tag{1.28}$$

Proof: The proof results from (1.22) to (1.24). ∎

Thus we have shown that if $\bar{z}$ is a solution of the equation (1.10) satisfying (1.11), then $z = S_\theta \bar{z}$ with appropriate choice of θ is an approximate solution of order (μ,ν,σ) with $\sigma = \nu$ of the linearized equation (1.18); that is, there exist a constant $M > 0$ and a vector z that satisfy relations (1.25) to (1.27).

Now put $z = x + h$. Then it follows from (1.27) that h is a solution of the equation

$$Lh + F'(x)h + Px + y = 0 \tag{1.29}$$

and we get from (1.25), (1.26), and (1.13),

$$\|h\|_0 \leq C_0(\|Px\|_0 + MQ^{-\mu}K^{\sigma}) \tag{1.30}$$

$$\|h\|_p \leq \|x\|_p + MQK^{\nu} \tag{1.31}$$

where $\sigma = \nu$ (throughout this section).

1.4 LEMMA Let $x \in V_0$ with $\|x\|_p < K$, and let

$$q_0C_0 = \bar{q} \quad \text{and} \quad \bar{q} < \frac{q}{2} < \frac{1}{2} \tag{1.32}$$

If h is a solution of the equation (1.29), then there exists $0 < \varepsilon \leq 1$ such that

$$\|P(x + \varepsilon h) - (1 - \varepsilon)Px\|_0 \leq \varepsilon q\|Px\|_0 \tag{1.33}$$

and

$$\|h\|_0 \leq C_0(1 + \bar{q})\|Px\|_0 \tag{1.34}$$

Proof: The proof is exactly the same as in Lemma 1.1, Chapter 1. ∎

1.5 LEMMA For $x_n \in V_0$ let

$$\|x_n\|_p < A\exp(\alpha(1 - q)t_n) = K_n \tag{1.35}$$

where $t_n > 0$, and let α, A, $\bar{q}$, and q be such that

$$\max([\mu(1 - \nu) - \nu]^{-1}, (1 - q)^{-1}) < \alpha \tag{1.36}$$

$$\mu(1 - \nu) - \nu > 0 \tag{1.37}$$

$$M^{1+1/\mu}(\bar{q}p_0)^{-1/\mu} < A^{1-\nu-\nu/\mu}[\alpha(1 - q) - 1] \qquad \alpha > 1 \tag{1.38}$$

where $p_0 = \|Px_0\|_0$, and q, q_0 satisfy (1.13), Chapter 1. Then there exists a number $Q = Q_n$ such that

$$MQ_n^{-\mu}K_n^{\nu} < \bar{q}p_0 \exp(1(1 - q)t_n) \tag{1.39}$$

and

$$\|x_n + \varepsilon h_n\|_p < A \exp(\alpha(1 - q)(t_n + \varepsilon)) \tag{1.40}$$

for all $0 < \varepsilon \leq 1$, where h_n is a solution of the equation (1.29) with $x = x_n$, $y = y_n$, provided that relations (1.30) and (1.31) hold with $x = x_n$, $h = h_n$, $K = K_n$, $Q = Q_n$.

Proof: The proof of Lemma 1.2, Chapter 1, carries over with $\|\cdot\|_X$, $\|\cdot\|_Z$ replaced by $\|\cdot\|_0$, $\|\cdot\|_p$, respectively. ∎

The following is the key lemma for the degree $0 < \bar{k} < 1$ of elliptic regularization.

1.6 LEMMA For $0 < \bar{\nu} < 1$ there exist μ and ν which satisfy (1.21) and such that for $K > 1$ and $Q > 1$ one can find $0 < \eta < 1$ and $\theta > 1$, which satisfy (1.22) and (1.23) if

$$0 < \bar{k} < 1 - \bar{\nu} \tag{1.40a}$$

Proof: Let $0 < \eta < \theta^{-m}$. Then (1.23) can be replaced by

$$\theta^{-m}\eta^{-\bar{k}}K^{\bar{\nu}} < Q^{-\mu}K^{\nu} \tag{1.41a}$$

Hence

$$\theta^{-m}Q^{\mu}K^{-(\nu-\bar{\nu})} < \eta^{\bar{k}} < \theta^{-m\bar{k}} \tag{1.42a}$$

and, by (1.22),

$$\theta^{p-m_2}Q^{-1}K^{-(\nu-\bar{\nu})} < \eta^{\bar{k}} < \theta^{-m\bar{k}} \tag{1.43a}$$

The inequalities (1.42a) and (1.43a) yield, respectively,

$$\theta^{-(1-\bar{k})m} < Q^{-\mu}K^{\nu-\bar{\nu}} \quad \text{or} \quad [Q^{\mu}K^{-(\nu-\bar{\nu})}]^{1/(1-\bar{k})m} < \theta$$

and

$$\theta^{p-m_2+m\bar{k}} < QK^{\nu-\bar{\nu}} \quad \text{or} \quad \theta < [QK^{(\nu-\bar{\nu})}]^{1/(p-m_2+mk)}$$

Hence

$$Q^{\mu/(1-\bar{k})m-1/(p-m_2+m\bar{k})} < K^{(\nu-\bar{\nu})[1/(p-m_2+m\bar{k})+1/(1-\bar{k})m]} \tag{1.44a}$$

provided

$$\frac{\mu}{(1-\bar{k})m} - \frac{1}{p-m_2+m\bar{k}} > 0$$

which yields

$$\mu > \frac{(1-\bar{k})m}{p-m_2+m\bar{k}} \tag{1.45a}$$

We get from (1.44a)

$$Q < K^{(\nu-\bar{\nu})\xi} \tag{1.46a}$$

where

$$\xi = \frac{p-m_2+m}{\mu(p-m_2+m\bar{k}) - (1-\bar{k})m}$$

But (1.46a) is compatible with (1.25) and (1.26), Chapter 1, if

$$1-\nu < (\nu-\bar{\nu})\xi \qquad \text{or} \qquad \frac{1+\bar{\nu}\xi}{1+\xi} < \nu$$

which together with $\nu < \mu/(1+\mu)$ by (1.21) yield

$$(1+\xi)^{-1}(1+\bar{\nu}\xi) < \frac{\mu}{1+\mu}$$

Hence

$$1 < [(1-\bar{\nu})\mu - \bar{\nu}]\xi = \frac{[(1-\bar{\nu})\mu - \bar{\nu}](p-m_2+m)}{\mu(p-m_2+m\bar{k}) - (1-\bar{k})m}$$

But this implies

$$\mu(p-m_2+m\bar{k}) - (1-\bar{k})m < [(1-\bar{\nu})\mu - \bar{\nu}](p-m_2+m)$$

or

$$\mu[\bar{\nu}(p-m_2) - (1-\bar{\nu}-\bar{k})m] < [(1-\bar{\nu}-\bar{k})m - \bar{\nu}(p-m_2)]$$

which is satisfied if

$$(1 - \bar{\nu} - \bar{k})m - \bar{\nu}(p - m_2) > 0$$

or

$$\bar{\nu} < \frac{(1 - \bar{k})m}{p - m_2 + m} \tag{1.47a}$$

Put $\bar{\nu} = (\bar{p} - \rho)/(p - \rho)$ with $0 < \rho < \bar{p}$ to be determined. Then (1.47a) holds true if

$$\frac{\bar{p}(p - m_2 + m) - p(1 - \bar{k})m}{p - m_2 + \bar{k}m} < \rho < \bar{p} \tag{1.48a}$$

Such ρ exists, since obviously

$$\bar{p}(p - m_2 + m) - p(1 - \bar{k})m < \bar{p}[p - m_2 + \bar{k}m]$$

is satisfied because $\bar{p} < p$ by assumption.

Furthermore, the constant α involved in Theorem 1.1, Chapter 1, satisfies conditions (1.19) and (1.33), Chapter 1, where $\sigma = \nu$ and $\bar{s} = s/p$, that is,

$$[\mu(1 - \nu) - \nu]^{-1} < \alpha < \frac{p - s}{s} \tag{1.49a}$$

Hence

$$\frac{s}{p - s} < \mu(1 - \nu) - \nu$$

or

$$s < p\,\frac{\mu(1 - \nu) - \nu}{(1 + \mu)(1 - \nu)} \tag{1.50a}$$

where $\rho \le s < \bar{p}$. Now consider the inequality

$$\bar{p} < p\,\frac{\mu(1 - \nu) - \nu}{(1 + \mu)(1 - \nu)} \tag{1.51a}$$

which yields

$$(p - \bar{p})(1 + \mu)\nu < (p - \bar{p})\mu - \bar{p}$$

provided

$$\mu > \frac{\bar{p}}{p - \bar{p}} \tag{1.52a}$$

Hence we get

$$\bar{\nu} = \frac{\bar{p} - \rho}{p - \rho} < \nu < \frac{(p - \bar{p})\mu - \bar{p}}{(p - \bar{p})(1 + \mu)} \tag{1.53a}$$

The inequalities (1.53a) have a solution for ν if ρ is such that

$$\bar{p}(2p - \bar{p}) - (p - \bar{p})^2\mu < \rho p$$

But $\rho < \bar{p}$, by assumption. Hence we get

$$\frac{\bar{p}(2p - \bar{p}) - (p - \bar{p})^2\mu}{p} < \rho < \bar{p} \tag{1.54a}$$

But ρ satisfying (1.54a) exists if

$$\bar{p}(2p - \bar{p}) - (p - \bar{p})^2\mu < p\bar{p}$$

which is equivalent to (1.52a). Consequently, the following is the order of operations. First choose μ to satisfy (1.45a) and (1.52a). Then choose ρ to satisfy (1.48a) and (1.54a). Next ν can be chosen to satisfy (1.53a) and (1.51a). In addition to (1.49a) α is also subject to

$$[1 - q]^{-1} < \alpha < \frac{p - s}{s} \tag{1.55a}$$

by virtue of (1.19), Chapter 1, which yields

$$s < \frac{p(1 - q)}{2 - q} \tag{1.56a}$$

Therefore (1.50a) and (1.56a) hold for s such that

$$s < \min[\bar{p}, p(1 - q)/(2 - q)] \tag{1.57a}$$

by virtue of (1.51a). Now one can choose α such that

$$\max[(1 - q)^{-1}, [\mu(1 - \nu) - \nu]^{-1}] < \alpha < \frac{p - s}{s} \tag{1.58a}$$

with s determined by (1.57a). ∎

1.1 REMARK From the standpoint of applications the condition $s > m_2$, in addition to (1.48a) and (1.54a), is rather important, especially in case of using evolution operators for the linearized equations.

In the same way as in Section 1, Chapter 1, we construct the following method of contractor directions.

$$x_{n+1} = x_n + \varepsilon_n h_n \qquad t_{n+1} = t_n + \varepsilon_n \qquad t_0 = 0 \tag{1.41}$$

where h_n is a solution of the equation (1.29) with $x = x_n$, $y = y_n$, and with induction assumptions

$$\|x_n\|_p < A \exp(\alpha(1 - q)t_n) = K_n \tag{1.42}$$

$$\|Px_n\|_0 \le \|Px_0\|_0 \exp(-(1 - q)t_n \tag{1.43}$$

where α, A, and q are subject to

$$\max([\mu(1 - \nu) - \nu]^{-1}, (1 - q)^{-1}) < \alpha < \frac{p - s}{s} \tag{1.44}$$

and (1.38) so that $Q = Q_n$ can be found to satisfy (1.25) and (1.26), Chapter 1. The choice of $\{\varepsilon_n\}$ is exactly the same as in Section 1, Chapter 1, with $\|\cdot\|$ replaced by $\|\cdot\|_0$. The estimate (1.34), Chapter 1, is still valid with $\|\cdot\|_Y$ replaced by $\|\cdot\|_s$, where s is chosen so as to satisfy

$$\max([\mu(1 - \nu) - \nu]^{-1}, (1 - q)^{-1}) < \frac{p - s}{s} \tag{1.45}$$

which is a consequence of (1.33), Chapter 1, and one has to put $\bar{s} = s/p$ in Section 1, Chapter 1.

1.1 THEOREM In addition to the assumptions (A_0) to (A_4), suppose that conditions (1.13) and (1.36) in Chapter 1, and (1.44) and (1.38) are satisfied. Then the equation (1.1) has a solution x such that $\|x_n - x\|_s \to 0$ as $n \to \infty$, where $\{x_n\} \subset V_0$ is determined by (1.41) and $\|x - x_0\|_s < r_0$.

Proof: The proof of Theorem 1.1, Chapter 1, carries over. ∎

1.2 REMARK Assumption (A_3) can be replaced by a weaker one.

(A_3') $x \in V_0$ with $\|x\|_{\bar{p}} \leq K$ and $\|g\|_{m_2} \leq K$ imply that the equation

$$Lz + F'(x)z + \eta Lz + g = 0 \tag{1.46}$$

has a solution $\bar{z}$ such that $\|\bar{z}\|_j < CK$, for some $C > 0$ and $j \geq m_2$. If $j > m_2$, then m_2 in (1.21), (1.22), and (1.9), but not in (1.20), should be replaced by j. This can be useful in applications. The other assumptions in (A_3) remain unchanged.

1.3 REMARK Let $\{X_j\}$ with $0 \leq j \leq p$ be a scale of Hilbert spaces. Suppose that the equation (1.10) with $L = 0$ has a solution, and

$$0 \leq (F'(x)z, Lz)_{m_2} \quad \text{and} \quad \|z\|_{m_2} \leq \|Lz\|_{m_2} \tag{1.47}$$

Then the inequality (1.11) holds with $\bar{k} = 1$.

Proof: We have by (1.47)

$$\|L\bar{z}\|_{m_2}^2 \leq (\eta^{-1}g, L\bar{z})_{m_2} \leq \frac{1}{2}\eta^{-2}\|g\|_{m_2}^2 + \frac{1}{2}\|L\bar{z}\|_{m_2}^2$$

$$g = Fx - F'(x)x + fx$$

Hence

$$\|\bar{z}\|_{m_2}^2 \leq \eta^{-2}\|g\|_{m_2}^2 \quad \blacksquare \tag{1.48}$$

1.4 REMARK The inequality (1.48) still holds if the condition (1.47) is replaced by the following.

$$0 \leq (F'(x)z, z)_{m_2} \quad \text{and} \quad \|z\|_{m_2}^2 \leq (Lz, z)_{m_2} \tag{1.49}$$

Proof: We have

$$\|\bar{z}\|_{m_2}^2 \leq (\eta^{-1}g, \bar{z})_{m_2} \leq 2^{-1}\eta^{-2}\|g\|_{m_2}^2 + 2^{-1}\|\bar{z}\|_{m_2}^2$$

Hence (1.48) follows. ■

1.5 REMARK Suppose that (A_4) is satisfied with $L = 0$, and the condition (1.47) is replaced by the following.

$$\|z\|^2_{m_2} \le (F'(x)z,z)_{m_2} + K_0^2\|z\|_0^2 \quad \text{and} \quad 0 \le (Lz,z)_{m_2} \tag{1.50}$$

Then

$$\|\bar{z}\|_{m_2} \le C\|g\|_{m_2} \tag{1.51}$$

is satisfied for some $C > 0$.

Proof: We have

$$(\eta L\bar{z},\bar{z})_{m_2} + (F'(x)\bar{z},\bar{z})_{m_2} = (g,\bar{z})_{m_2}$$

Hence, we get

$$\|\bar{z}\|^2_{m_2} \le 2^{-1}\|g\|^2_{m_2} + 2^{-1}\|\bar{z}\|^2_{m_2} + K_0^2\|\bar{z}\|_0^2$$

and (1.51) follows by (A_4). ■

1.6 REMARK Suppose that

$$\|z\|^2_{j+i} \le (F'(x)z,Lz)_j + K_0^2\|z\|_0^2 \quad \text{and} \quad \|Lz\|_j \le \|z\|_{j+i} \tag{1.52}$$

Then

$$\|\bar{z}\|_{j+i} \le C\|g\|_{j+i} \tag{1.53}$$

holds for some $C > 0$.

Proof: We have

$$\begin{aligned}\eta\|L\bar{z}\|_j^2 + (F'(x)\bar{z},Lz)_j + K_0^2\|\bar{z}\|_0^2 &= (g,L\bar{z})_j + K_0^2\|\bar{z}\|_0^2 \\ &\le 2^{-1}\|g\|_j^2 + 2^{-1}\|\bar{z}\|^2_{j+1} + K_0^2\|\bar{z}\|_0^2\end{aligned}$$

Hence, (1.53) follows by (A_4). ■

1.7 REMARK Definition 1.2, Chapter 1, can also be applied to this section so that the results and the proofs will remain valid. To this end the following lemma is needed to replace Lemma 1.6.

1.6a LEMMA For $0 < \bar{\nu} < 1$ and $0 < \bar{k} < 1$, there exist ν and σ with

$\sigma > 0$ and $\bar{\nu} < \nu \le 1$ and such that for $K > 1$ one can find $0 < \eta < 1$ and $\Theta > 1$, which satisfy the following conditions:

$$\Theta^{p-m_2}\eta^{-\bar{k}}K^{\bar{\nu}} < K^{\nu} \tag{1.54}$$

$$(\Theta^{-m} + \eta)\eta^{-\bar{k}}K^{\bar{\nu}} < K^{-\sigma} \tag{1.55}$$

Proof: Let $0 < \eta < \Theta^{-m}$. Then (1.55) can be replaced by

$$\Theta^{-m}\eta^{-\bar{k}}K^{\bar{\nu}} < K^{-\sigma} \tag{1.56}$$

Hence

$$\Theta^{-m}K^{\bar{\nu}+\sigma} < \eta^{\bar{k}} < \Theta^{-m\bar{k}} \tag{1.57}$$

and by (1.54), we get

$$\Theta^{p-m_2}K^{-(\nu-\bar{\nu})} < \eta^{\bar{k}} < \Theta^{-m\bar{k}} \tag{1.58}$$

The inequalities (1.57) and (1.58) yield, respectively,

$$K^{\bar{\nu}+\sigma} < \Theta^{m(1-\bar{k})} \quad \text{or} \quad K^{(\bar{\nu}+\sigma)/m(1-\bar{k})} < \Theta$$

and

$$\Theta^{p-m_2+m\bar{k}} < K^{\nu-\bar{\nu}} \quad \text{or} \quad \Theta < K^{(\nu-\bar{\nu})/(p-m_2+m\bar{k})}$$

Hence

$$K^{(\bar{\nu}+\sigma)/m(1-\bar{k})} < \Theta < K^{(\nu-\bar{\nu})/(p-m_2+m\bar{k})}$$

and

$$\frac{\bar{\nu} + \sigma}{m(1 - \bar{k})} < \frac{\nu - \bar{\nu}}{p - m_2 + m\bar{k}}$$

or

$$\bar{\nu} + \sigma < (\nu - \bar{\nu})\xi \qquad \xi = \frac{m(1 - \bar{k})}{p - m_2 + m\bar{k}}$$

which is satisfied if

$$[\bar{\nu}(1 + \xi) + \sigma]/\xi < \nu \le 1 \tag{1.59}$$

and consequently,

$$\bar{\nu}(1 + \xi) + \sigma < \xi$$

Hence,

$$\sigma < \xi(1 - \bar{\nu}) - \bar{\nu} \tag{1.60}$$

Since $\sigma > 0$, it follows from (1.60) that

$$\bar{\nu} < \frac{\xi}{1 + \xi} = \frac{m(1 - \bar{k})}{p - m_2 + m} = c < 1$$

Put

$$\bar{\nu} = \frac{\bar{p} - \rho}{p - \rho} \tag{1.61}$$

with ρ to be determined from (1.62). Hence, $(\bar{p} - \rho)/(p - \rho) < c$, and we choose ρ to satisfy

$$\frac{\bar{p} - cp}{1 - c} < \rho < \bar{p} \tag{1.62}$$

Such $\rho > 0$ exists, since $\bar{p} < p$. ∎

It is easily seen that

$$\frac{\bar{p}}{p} > \frac{m}{p - m_2 + m} > c$$

Consequently, the following is the order of operations. First choose ρ to satisfy (1.62). Next, determine $\bar{\nu}$ from (1.61). Then ν and σ can be chosen to satisfy (1.59) and (1.60), respectively.

2. GLIM-II IN A SCALE OF BANACH SPACES

Let $\{X_j\}$ with $0 \leq j \leq p$ be a scale of Banach spaces with increasing norms such that $i < j$ implies $X_j \subset X_i$ and $\|x\|_j \geq \|x\|_i$, and let $0 < m_1 < m_2 \leq \bar{p} < p$. We make the following assumptions.

(A_0) We assume that there exists a one-parameter family of linear smoothing operators S_θ, $\theta \geq 1$, which satisfies the same conditions as in Section 1.

Let $W_0 \subset X_s$ be an open ball with center $x_0 \in X_p$ and radius $r_0 > 0$, s to be determined. Put $V_0 = W_0 \cap X_p$ and let V_s be the $\|\cdot\|_s$-closure of V_0.

Consider the nonlinear operator equation

$$Px \equiv \dot{A}x + Fx + fx = 0 \tag{2.1}$$

where P is closed on V_s and assumes its values in the same scale of Banach spaces provided that $\dot{A} \neq 0$, $\dot{A}$ being linear and nonbounded. If the domain of P is $D \subset X_s$, then we put $V_0 = D \cap W_0 \cap X_p$.

We make the following assumptions.

(A_1) F is differentiable, i.e., for each $x \in V_0$, there exists a linear operator $F'(x)$ such that

$$\varepsilon^{-1}\|F(x + \varepsilon h) - Fx - \varepsilon F'(x)h\|_0 \to 0 \qquad \text{as } \varepsilon \to 0+ \tag{2.2}$$

There exists a constant $C > 0$ such that

$$\|F'(x)h\|_{m_1} \leq C\|h\|_{m_2} \tag{2.3}$$

$$\|F'(x)h\|_0 \leq C\|h\|_{m_1} \tag{2.4}$$

for all $x \in V_0$.

There exists a constant $C_1 > 0$ such that

$$\|F(x + h) - Fx - F'(x)h\|_0 \leq C_1\|h\|_0\|h\|_s \tag{2.5}$$

for all $x \in V_0$. There exists a constant $C > 0$ such that

$$\|Fx - F'(x)x + fx\|_{m_2} \leq C\|x\|_{\bar{p}} \tag{2.6}$$

and

$$\|fx\|_{m_2} \leq C\|x\|_{\bar{p}} \tag{2.7}$$

(A_2) There exists a constant q_0 such that

$$\|f(x + h) - fx\|_0 \leq q_0\|h\|_0 \tag{2.8}$$

for all $x \in V_0$.

(A_3) There exists a linear (regularizing) operator $L = L(\eta)$ such that

$$\|Lx\|_0 \le C\|x\|_{m_2} \tag{2.9}$$

for some constant $C > 0$.

The linearized equation

$$\dot{A}z + F'(x)zL + \eta Lz + Fx - F'(x)x + fx = 0 \tag{2.10}$$

with small $0 < |\eta| < 1$ has a solution $\bar{z}$ such that

$$\|\bar{z}\|_{m_2} \le C|\eta|^{-\bar{k}}\|Fx - F'(x)x + fx\|_{m_2} \tag{2.11}$$

for some $\bar{k}$ to be determined and $C > 0$; $\bar{k}$ is called the degree of elliptic regularization.

(A_4) For $x \in V_0$, if h is a solution of the equation

$$\dot{A}h + F'(x)h + g = 0 \tag{2.12}$$

then

$$\|h\|_0 \le C_0\|g\|_0 \tag{2.13}$$

for some constant $C_0 > 0$. In order to simplify the notation we assume that $0 < \eta < 1$.

For $x \in V_0$, let $\bar{z}$ be a solution of the equation (2.10) satisfying (2.11). Then the following estimates hold.

2.1 LEMMA The following relations hold for $x \in V_0$ with $\|x\|_p < K$, $K > 1$, and $0 < \eta < 1$.

$$\|\dot{A}(I - S_\theta)\bar{z}\|_0 \le M_1(\theta^{-m_1} + \eta)\eta^{-\bar{k}}K^{\bar{\nu}} \tag{2.14}$$

$$\|F'(x)(I - S_\theta)\bar{z}\|_0 \le M_2\theta^{-(m_2-m_1)}\eta^{-\bar{k}}K^{\bar{\nu}} \tag{2.15}$$

$$\|z\|_p = \|S_\theta\bar{z}\|_p \le M_3\theta^{p-m_2}\eta^{-\bar{k}}K^{\bar{\nu}} \tag{2.16}$$

for some M_1, M_2, M_3, and $z = S_\theta\bar{z}$, where $\bar{\nu} = (\bar{p} - \rho)/(p - \rho)$.

Proof: For the proof see Lemma 1.1. ∎

Consider the linearized equation

$$\dot{A}z + F'(x)z + Fx - F'(x)x + fx = 0 \tag{2.17}$$

and put $z = S_\theta \bar{z}$, where $\bar{z}$ is a solution of the linearized equation (2.10) satisfying relation (2.11).

2.2 LEMMA The following estimate holds:

$$\|\dot{A}z + F'(x)z + Fx - F'(x)x + fx\|_0 \le M_4(\theta^{-m} + \eta)\nu^{-\bar{k}}K^{\bar{\nu}} \tag{2.18}$$

for $x \in V_0$ with $\|x\|_p < K$, and some $M_4 > 0$, where

$$m = \min(m_1, m_2 - m_1) \tag{2.19}$$

Proof: For the proof see Lemma 1.2. ∎

Now suppose there exist ν,μ with

$$\bar{\nu} < \nu < 1 \quad \text{and} \quad \mu(1 - \nu) - \nu > 0 \tag{2.20}$$

such that for $Q > 1$, $K > 1$, one can find $0 < \eta < 1$ and $\Theta > 1$ to satisfy

$$\Theta^{p-m_2}\eta^{-\bar{k}}K^{\bar{\nu}} < QK^{\nu} \tag{2.21}$$

$$(\Theta^{-m} + \eta)\eta^{-\bar{k}}K^{\bar{\nu}} < Q^{-\mu}K^{\nu} \tag{2.22}$$

Then (2.18) can be written as

$$\|\dot{A}z + F'(x)z + Fx - F'(x)x + fx\|_0 \le 2M_4Q^{-\mu}K^{\nu} \tag{2.23}$$

and

$$\|z\|_p = \|S_\theta\bar{z}\|_p \le M_3QK^{\nu} \tag{2.24}$$

2.3 LEMMA The linearized equations (2.17) admit approximate solutions of order (μ,ν,σ) in the sense of Definition 1.1, with $\sigma = \nu$.

Proof: The proof follows from (2.24) and (2.23). ∎

It is clear that Lemma 2.3 holds under the assumption that $0 < \eta < 1$ and $\Theta > 1$ satisfy (2.21) and (2.22), with μ,ν to be determined. Thus we have shown that if $\bar{z}$ is a solution of the equation (2.10) satisfying (2.11), then $z = S_\theta \bar{z}$ is an approximate solution of order (μ,ν,σ) with $\sigma = \nu$ of the linearized equation (2.17); that is, there exist a constant $M > 0$ and vectors $z = S_\theta \bar{z}$, y, such that $x \in V_0$ with $\|x\|_p < K$, $K > 1$, $Q > 1$ imply

$$\dot{A}z + F'(x)z + Fx - F'(x)x + fx + y = 0 \tag{2.25}$$

$$\|z\|_p \le MQK^\nu \tag{2.26}$$

$$\|y\|_0 \le MQ^{-\mu}K^\nu \tag{2.27}$$

Now put $z = x + h$. Then it follows from (2.25) that h is a solution of the equation

$$\dot{A}h + F'(x)h + Px + y = 0 \tag{2.28}$$

and we get from (2.26) to (2.28) and (2.13),

$$\|h\|_0 \le C_0(\|Px\|_0 + MQK^\sigma) \tag{2.29}$$

$$\|h\|_p \le \|x\|_p + MQK^\nu \tag{2.30}$$

where $\sigma = \nu$.

2.4 LEMMA Let $x \in V_0$ with $\|x\|_p < K$, and let condition (1.13) be satisfied. Suppose that h is a solution of the equation (2.28). If $K > 1$ and $Q > 1$ are such that

$$\|y\|_0 \le MQ^{-\mu}K^\nu \le \bar{q}\|Px\|_0 \tag{2.31}$$

then

$$\|P(x + h)\| \le q\|Px\| \tag{2.32}$$

$$\|h\|_0 \le C_0(1 + \bar{q})\|Px\|_0 \tag{2.33}$$

provided that

$$(c_1 r_0 + q_0)C_0(1 + 2\bar{q}) < q < 1 \qquad \bar{q} < q \tag{2.34}$$

Proof: The proof is exactly the same as in Lemma 2.1, Chapter 1. However, relations (2.21), (2.22) impose additional restrictions on Q. See Lemma 2.5a and the proof of Lemma 1.6, where

$$Q < K^{(\nu-\bar{\nu})\xi} \qquad \xi = \frac{p - m_2 + m}{\mu(p - m_2 + mk) - (1 - \bar{k})m} \qquad \blacksquare \tag{2.35}$$

2.5 LEMMA For $x_n \in V_0$, let

$$\|x_n\|_p < Aq^{-\alpha n} = K_n \tag{2.36}$$

$$\|Px_n\|_0 \leq p_0 q^n \tag{2.37}$$

where $p_0 = \|Px_0\|_0$, and $\alpha > 1$, A are chosen so as to satisfy relations (2.20) and (2.21), Chapter 1, with $\sigma = \nu$. Then there exists a number $Q = Q_n$ such that

$$MQ_n^{-\mu}K^{\nu} \leq \bar{q}p_0q^n \tag{2.38}$$

and

$$\|x_n + h_n\|_p < AQ^{-\alpha(n+1)} = K_{n+1} \tag{2.39}$$

where h_n is a solution of the equation (2.28) with $x = x_n$, i.e.,

$$\dot{A}h_n + F'(x_n)h_n + Px_n + y_n = 0 \tag{2.40}$$

which satisfies relations (2.29) with $\sigma = \nu$ and (2.30).

Proof: The proof of Lemma 2.2, Chapter 2, carries over. ■

2.5a LEMMA There exist ν with $\bar{\nu} < \nu < 1$ and μ, which satisfies (2.20), and such that for $K > 1$ and $Q > 1$ one can find $0 < \eta < 1$ and $\Theta > 1$, which satisfy (2.21) and (2.22) provided

$$0 < \bar{k} < 1 \tag{2.41}$$

$$\bar{\nu} = \frac{\bar{p} - \rho}{p - \rho} \tag{2.42}$$

with ρ as in the proof of Lemma 1.6.

$$\mu > \bar{p}/(p - \bar{p}) \tag{2.43}$$

Proof: For the proof see Lemma 1.6. ■

Notice that m_2 in (2.21) and (2.9) can be replaced by $\bar{p}$. Thus μ can be found to satisfy (2.20), and consequently both conditions (1.26) and (2.35) can be satisfied.

We shall now construct the following iterative method. With x_0 as above, assume that $x_0, x_1, \ldots, x_n$ are known and satisfy the relations (2.36) and (2.37) of Lemma 2.5. Then put

$$x_{n+1} = z_n = x_n + h_n \tag{2.44}$$

where z_n and h_n are solutions of the equations (2.25) with $x = x_n$, $y = y_n$, and (2.28), respectively, i.e., z_n is an approximate solution of the linearized equation (2.17). It results from (2.39) that relation (2.36) holds true for n + 1, and relation (2.32) with $x = x_n$, $h = h_n$, implies that condition (2.37) is satisfied for n + 1.

2.6 LEMMA The following estimate holds.

$$\sum_{n=0}^{\infty} \|h_n\|_s \le \frac{N}{1 - q^{\beta}} \tag{2.45}$$

where $N = C[C_0(1 + \bar{q})p_0]^{1-\bar{s}}[A(q^{-\alpha} - 2)]^{\bar{s}}$, $\beta = [1 - \bar{s}(1 + \alpha)] > 0$, with $\bar{q}$ and q as in (2.34) and $\bar{s} = s/p$ such that

$$[\mu(1 - \nu) - \nu]^{-1} < \frac{1 - \bar{s}}{\bar{s}} = \frac{p - s}{s} \tag{2.46}$$

Proof: The proof is the same as for Lemma 2.3, Chapter 1. ■

Note that $\beta > 0$ implies $\alpha < (p - s)/s$ so that

$$[\mu(1 - \nu) - \nu]^{-1} < \alpha < \frac{p - s}{s} \tag{2.47}$$

by (1.18) and (1.19), Chapter 1, with $\sigma = \nu$.

2.1 THEOREM In addition to assumptions (A_0) to (A_4), suppose that α satisfies (2.47) and A satisfies (1.20), Chapter 1. If

$$N(1 - q^{\beta}) < r_0$$

then equation (2.1) has a solution x such that

$$\|x_n - x\|_s \to 0 \qquad \text{as } n \to \infty$$

and $\{x_n\} \in V_0$.

Proof: The proof of Theorem 2.1, Chapter 1, carries over. ■

2.1 REMARK Definition 1.2 of approximate linearization (see Remark 1.1, Chapter 1) can also be applied to this section. Then Lemma 2.5a should be replaced by Lemma 1.6a.

3. GLIM-III IN A SCALE OF BANACH SPACES

Let $\{X_j\}$ with $0 \le j \le p$ be a scale of Banach spaces with increasing norms such that $i < j$ implies $X_j \subset X_i$ and $\|x\|_j \ge \|x\|_i$, and let $0 < m_1 < m_2 < \bar{p} < p$. We make the following assumptions.

(A_0) We assume that there exists a one-parameter family of linear smoothing operators S_θ, $\theta \ge 1$, which satisfy the same conditions as in Section 1.

Let $W_0 \subset X_s$ be an open ball with center $x_0 \in X_p$ and radius $r_0 > 0$, s to be determined. Put $V_0 = W_0 \cap X_p$ and let V_s be the $\|\cdot\|_s$-closure of V_0.

Consider the nonlinear operator equation

$$Px \equiv Ax + Fx = 0 \tag{3.1}$$

where P is closed on V_s and assumes its values in the same scale of Banach spaces provided that $A \ne 0$. If the domain of P is $D \subset X_s$, then we put $V_0 = D \cap W_0 \cap X_p$.

We make the following assumptions.

(A_1) F is differentiable, i.e., for each $x \subset V_0$, there exists a linear operator $F'(x)$ such that

$$\varepsilon^{-1}\|F(x + \varepsilon h) - Fx - \varepsilon F'(x)h\|_0 \to 0 \qquad \text{as } \varepsilon \to 0+ \tag{3.2}$$

and

$$\|F(x + h) - Fx - F'(x)h\|_0 \le \bar{M}\|h\|_0^{2-\beta}\|h\|_p^{\beta} \tag{3.3}$$

for some constant $\bar{M} > 0$, $0 \le \beta < 1$, and all $x \in V_0$.

There exists a constant $C > 0$ such that

$$\|F'(x)h\|_{m_1} \le C\|h\|_{m_2} \qquad \|Fx - F'(x)x\|_{m_2} \le C\|x\|_{\bar{p}} \tag{3.4}$$

$$\|F'(x)h\|_0 \le C\|h\|_{m_1} \tag{3.5}$$

for all $x \in V_0$.

(A_2) For $x \in V_0$, if h is a solution of the equation

$$Ah + F'(x)h + g = 0 \tag{3.6}$$

then

$$\|h\|_0 \le C_0\|g\|_0 \tag{3.7}$$

for some constant $C_0 > 0$.

(A_3) There exists a linear (regularizing) operator $L = L(\eta)$ such that

$$\|Lx\|_0 \le C\|x\|_{m_2} \tag{3.8}$$

for some constant $C > 0$.

The linearized equation

$$Az + F'(x)z + \eta Lz + Fx - F'(x)x = 0 \tag{3.9}$$

with small $0 < |\eta| < 1$ has a solution $\bar{z}$ such that

$$\|\bar{z}\|_i \le C|\eta|^{-\bar{k}}\|Fx - F'(x)x\|_{m_2} \qquad i = \bar{p} \text{ or } m_2 \tag{3.10}$$

for some $\bar{k}$ to be determined and $C > 0$. $\bar{k}$ is called the degree of elliptic regularization.

3.1 DEFINITION If assumption (A_3) is satisfied then we say that the equation (3.1) admits elliptic regularization of degree $\bar{k}$.

In order to simplify the notation we assume $0 < \eta < 1$.

For $x \in V_0$, let $\bar{z}$ be a solution of the equation (3.9) satisfying (3.10). Then the following estimates hold.

3.1 LEMMA The following estimates hold for $x \in V_0$ with $\|x\|_p < K$, $K > 1$.

$$\|A(I - S_\theta)\bar{z}\|_0 \le M_1(\theta^{-m_1} + \eta)\eta^{-\bar{k}}K^{\bar{\nu}} \tag{3.11}$$

$$\|F'(I - S_\theta)\bar{z}\|_0 \le M_2\theta^{-(m_2-m_1)}\eta^{-\bar{k}}K^{\bar{\nu}} \tag{3.12}$$

$$\|z\|_p \equiv \|S_\theta \bar{z}\|_p \le M_3\theta^{p-\bar{p}}\eta^{-\bar{k}}K^{\bar{\nu}} \tag{3.13}$$

for some M_1, M_2, $M_3 > 0$ and $z = S_\theta\bar{z}$, $\bar{\nu} = (\bar{p} - \rho)/(p - \rho)$ with ρ to be determined.

Proof: For the proof see Lemma 1.1. ■

Consider the linearized equation

$$Az + F'(x)z + Fx - F'(x)x = 0 \tag{3.14}$$

and put $z = S_\theta\bar{z}$, where $\bar{z}$ is a solution of the linearized equation (3.9) satisfying (3.10).

3.2 LEMMA The following estimate holds.

$$\|Az + F'(x)z + Fx - F'(x)x\|_0 \le M_4(\theta^{-m} + \eta)\eta^{-\bar{k}}K^{\bar{\nu}} \tag{3.15}$$

for $x \in V_0$ with $\|x\|_p < K$ and some $M_4 > 0$, where

$$m = \min(m_1, m_2 - m_1) \tag{3.16}$$

Proof: For the proof see Lemma 1.1. ■

Now put $i = \bar{p}$ in (3.10)

$$Q = \theta^{p-\bar{p}} \tag{3.17}$$

$$\mu = \frac{m}{p - \bar{p}} \tag{3.18}$$

Let $0 < \alpha < 1$, and choose $0 < \eta < 1$ so as to satisfy the relations

$$\eta^{-\bar{k}}K^{\bar{\nu}} < K^{1+\alpha} \qquad \text{and} \qquad \eta < Q^{-\mu} \tag{3.19}$$

Then relation (3.15) can be written as

$$\|Az + F'(x)z + Fx - F'(x)x\|_0 \le 2M_4Q^{-\mu}K^{1+\alpha} \tag{3.20}$$

since $\theta^{-m} = Q^{-\mu}$ by (3.17), (3.18), and (3.19).

3.3 LEMMA The linearized equation (3.14) admits approximate solutions of order (μ,ν,σ) with $\nu = \sigma = 1 + \alpha$ in the sense of Definition 1.1 with $X = X_0$, $Z = Z_p$.

Proof: The proof follows from (3.17) to (3.20), and (3.13). ∎

Thus we have shown that if $\bar{z}$ is a solution of the equation (3.9) satisfying (3.10), then $z = S_\theta\bar{z}$ is an approximate solution of the linearized equation (3.14) of order (μ,ν,σ) with $\nu = \sigma = 1 + \alpha$, that is, there exist a vector y and a constant $M > 0$ such that

$$Az + F'(x)z + Fx - F'(x)x + y = 0 \tag{3.21}$$

$$\|z\|_p = \|S_\theta\bar{z}\|_p \le MQK^{1+\alpha} \quad \text{and} \quad \|y\|_0 \le MQ^{-\mu}K^{1+\alpha} \tag{3.22}$$

provided that $x \in V_0$ with $\|x\|_p < K$ $(K > 1)$, and η in (3.9) is chosen so as to satisfy (3.19).

Now put $z = x + h$. Then it follows from (3.21) that h is a solution of the equation

$$Ah + F'(x)h + Px + y = 0 \tag{3.23}$$

We shall construct an iterative method in the same way as in Section 1. With x_0 as above put

$$x_{n+1} = z_n = x_n + h_n \qquad x_n \in V_0 \tag{3.24}$$

where z_n is a solution of equation (3.21) with $x = x_n$, $y = y_n$, and Q_n in (3.22) is chosen so as to satisfy the induction assumptions

$$\|x_n\|_p < K_n \tag{3.25}$$

$$\|Px_n\|_0 \le K_n^{-\lambda} \tag{3.26}$$

where $K_{n+1} = K_n^{\tau+\alpha}$, and λ, α, and τ satisfy the assumptions (A_4) of

Section 3, Chapter 1. In order to verify the induction assumptions (3.25) and (3.26), one has to solve the system of inequalities (3.26) to (3.28) in Chapter 1. In addition, relations (3.19) with $K = K_n$, $Q = Q_n$, and $\eta = \eta_n$ imply that

$$Q < K_n^{(1+\alpha-\bar{\nu})/\mu\bar{k}} \tag{3.27}$$

But (3.26), Chapter 1, implies that $Q < K_n^{\tau-1}$. Hence, relation (3.27) is satisfied if $\bar{k}$ is such that

$$\tau - 1 < \frac{1 + \alpha - \bar{\nu}}{\mu\bar{k}} \tag{3.28}$$

or if

$$\bar{k} < \frac{1 + \alpha - \bar{\nu}}{(\tau - 1)\mu} \tag{3.29}$$

Then $\bar{k} \geq 1$ is feasible if $1 + \alpha - \bar{\nu} > (\tau - 1)\mu$ or if s can be chosen so as to satisfy

$$\frac{\bar{p} - s}{p - s} < 1 + \alpha - (\tau - 1)\mu \tag{3.30}$$

But this is possible if

$$1 + \alpha > (\tau - 1)\mu \qquad \text{or} \qquad \mu < \frac{1 + \alpha}{\tau - 1}$$

Hence, by (3.11),

$$\frac{1 + \alpha_0}{1 - \alpha_0} < \frac{1 + \alpha}{\tau - 1}$$

which implies

$$\tau < \frac{(1 + \alpha)(1 - \alpha_0)}{1 + \alpha_0} + 1$$

Hence, by (3.9), Chapter 1, we get the relation

$$\frac{1 + \alpha(1 + \lambda) + \mu}{\mu - \lambda} < \frac{(1 + \alpha)(1 - \alpha_0)}{1 + \alpha_0} + 1$$

and consequently, we obtain

$$\frac{(1 + \lambda)(1 + \alpha_0)}{1 - \alpha_0} < \mu - \lambda$$

or

$$2\lambda < \mu(1 - \alpha_0) - (1 + \alpha_0) = (1 + \alpha_0)[\mu(1 - \alpha_0)/(1 + \alpha_0) - 1]$$

which is satisfied by virtue of (3.10), Chapter 1.

3.1 THEOREM In addition to assumptions (A_0) to (A_3), suppose that (A_4), Section 3, Chapter 1, is satisfied. Then there exists a constant $K_0(M,\beta,\mu,\lambda,\alpha) > 1$ such that

$$\|Px_0\|_0 < K_0 \qquad \|x_0\|_p < K_0$$

and

$$C(2C_0)^{1-s/p} \sum_{n=0}^{\infty} K_n^{-\delta} < r_0$$

hold, where $\delta = (1 - s/p)\lambda - s\tau/p > 0$, provided

$$\frac{s}{p} < \frac{\lambda}{\lambda + 2}$$

and $K_{n+1} = K_n^{\tau+\alpha}$ with τ,α as in (3.9), Chapter 1, then equation (3.1) has a solution x which is a limit of $\{x_n\}$, and

$$\|x_n - x\|_s \to 0 \qquad \text{as } n \to \infty$$

where $\{x_n\}$ is determined by (3.24).

Proof: The proof follows from that of Theorem 3.1, Chapter 1. ∎

Consider now the case where $\nu = \sigma = 1$ in Definition 3.1, Chapter 1. Here assumption (A_4) of Section 3, Chapter 1, is replaced by (A'_4), and we obtain the relations (3.27) to (3.29) with $\alpha = 0$. Thus $\bar{k}$ is restricted by (3.29) with $\alpha = 0$, and $\bar{k} < 1$. In fact, suppose that $\bar{k} \geq 1$. Then we get from (3.30) with $\alpha = 0$ that

$(\tau - 1)\mu < 1$ or $\tau < 1/\mu + 1$. Hence, we get by (3.36), Chapter 1, that

$$\frac{\mu + 1}{\mu - \lambda} < \tau < \frac{\mu + 1}{\mu}$$

which is a contradiction.

3.2 THEOREM Theorem 3.1 remains valid if (A_4) is replaced by (A'_4) and $\nu = \sigma = 1$ in Definition 1.1, and $K_{n+1} = K_n^\tau$, $n = 0, 1, \ldots$.

Proof: The proof is the same as that of Theorem 3.1, where the relations (3.26) to (3.28), Chapter 1, should be replaced by (3.37) to (3.39), Chapter 1. But (3.37), Chapter 1, implies that $Q < K_n^{\tau-1}$, and the choice of η_n from (3.19) yields (3.27) with $\alpha = 0$. Hence, (3.29) follows with $\alpha = 0$. ∎

It is clear that the same method of constructing approximate solutions of the linearized equations, which is based on smoothing operators combined with elliptic regularization, can also be applied to Moser's (1956) theorem 1.3.

We notice the advantage of GLIM-III with $0 < \alpha < 1$ in Theorem 3.1, which imposes a rather weaker restriction on the degree of elliptic regularization, i.e., $\bar{k} \geq 1$ is admissible.

3.1 REMARK Another feature of the method under investigation is indicated by the condition (3.10), which means in practice that regularization is required with no gain of derivatives.

3.2 REMARK The following is a sufficient condition for (3.10) to be satisfied if $A = 0$.

$$|\eta| \cdot \|w\|_{m_2} \leq \|\eta Lw + F'(x)w\|_{m_2} \qquad 0 < |\eta| < 1,\ x \in V_0$$

or in case of Sobolev-Hilbert spaces,

$$|\eta| \cdot (w,w)_{m_2} = |\eta| \cdot \|w\|_{m_2}^2 \leq (\eta Lw, w)_{m_2} + (F'(x)w,w)_{m_2}$$

or a stronger condition

$$\|w\|_{m_2}^2 \le (Lw,w)_{m_2} \qquad \text{and} \qquad (F'(x),w,w) \ge 0 \qquad x \in V_0$$

Then evidently the degree of elliptic regularization is $\bar{k} = 1$.

With μ determined by (3.18), Theorem 3.2 holds true under condition (3.29) (with $\alpha = 0$) imposed on $\bar{k}$. On the other hand, Lemma 1.6, which applies only to GLIM-I and GLIM-II, is valid for arbitrary $0 < \bar{k} < 1$. But a similar lemma can be proved for GLIM-III.

3.4 LEMMA There exists μ which satisfies conditions (1.22) and (1.23) with $\nu = 1$ if $0 < \bar{k} < 1$ is arbitrary.

Proof: Condition (1.46a) is satisfied with $\nu = 1$. Let $\tau - 1 < (1 - \bar{\nu})\xi$. Then

$$\frac{1 + \lambda\tau}{\mu} < \tau - 1 < (1 - \bar{\nu})\xi \tag{3.31}$$

by virtue of (3.37) and (3.39), Chapter 1. Substituting in (3.31) $\lambda < (\mu - 1)/2$ and $\tau < 2$ from (A'_4), Chapter 1, we get

$$1 < (1 - \bar{\nu})\xi \tag{3.32}$$

where $\xi = (p - m_2 + m)/[\mu(p - m_2 + m\bar{k}) - (1 - \bar{k})m]$. Hence we get, since $\mu > 1$,

$$1 < \mu < \frac{(1 - \bar{k})m + (1 - \bar{\nu})(p - m_2 + m)}{p - m_2 + m\bar{k}}$$

which implies

$$\bar{\nu}(p - m_2 + m) < 2(1 - \bar{k})m \tag{3.33}$$

Put $\bar{\nu} = (\bar{p} - \rho)/(p - \rho)$ in (3.33) with $0 < \rho < \bar{p}$. Then (3.33) yields

$$\bar{p}(p - m_2 + m) - p2(1 - \bar{k})m < \rho[p - m_2 + m - 2(1 - \bar{k})m]$$

Thus $0 < \rho < \bar{p}$ exists if

$$\bar{p}(p - m_2 + m) - p2(1 - \bar{k})m < \bar{p}[p - m_2 + m - 2(1 - \bar{k})m]$$

But this inequality reduces to

$$2(1 - \bar{k})m\bar{p} < 2(1 - \bar{k})mp$$

which is satisfied since $\bar{p} < p$. Note: put $i = m_2$ in (3.10).

It results from Lemma 3.4 that Theorem 3.2 is valid for arbitrary $0 < \bar{k} < 1$. ■

3.3 REMARK Definition 2.1 of approximate linearization (see Remark 1.1, Chapter 1) can also be applied to this section. To this end we utilize Remark 3.3, Chapter 1, where σ can be determined from Lemma 1.6a with $\nu = 1$.

3.4 REMARK It follows from the assumptions made in (A'_4), and (3.18), Section 3, Chapter 1, and from Lemma 3.4 that

$$\bar{s} = \frac{s}{p} < \frac{\lambda}{\lambda + 2} < 1 < \mu < \frac{(1 - \bar{k})m + (1 - \bar{\nu})(p - m_2 + m)}{p - m_2 + m\bar{k}} = d(p) \tag{3.34}$$

also holds for arbitrarily large p, and $d(p) > 1$. But this is not true if μ is determined as below, because the relation

$$\frac{m_2}{p} < \frac{m}{p - m_2} = \mu \qquad m = \min(m_1, m_2 - m_1)$$

is false. Hence, one can choose s such that $m_1 < s < m_2$, whereas the condition $m_2 < s < \bar{p}$ can be satisfied in case of (3.34).

Remark 3.4 is also valid for nonlinear evolution equations, and it is rather important from the standpoint of applications to nonlinear partial differential equations, especially if the C_0-semigroups theory is to be employed.

3

Nonlinear Equations in a Scale of Hilbert-Sobolev Spaces, I

This chapter presents a further development of a theory of global linearization iterative methods (GLIM). From the standpoint of applications to nonlinear partial differential equations (NPDE) the most important new idea that appears here is a certain method of elliptic regularization that yields the degree $\bar{k} = 1/2$. Here we consider only the case of smoothing operators combined with elliptic regularization. In some cases such an approach can be more effective than elliptic regularization without smoothing operators. Although the method under consideration still works in case of GLIM-II, it turns out that it may not be applicable in case of GLIM-III. This is due to the fact that in applications to NPDE, Moser's degree of approximate linearization can be smaller than one, i.e., $\mu < 1$, whereas GLIM-III requires $\mu > 1$.

There are also some other new features added here; e.g., the linear operator which appears in the linearized equation can be split into a sum of two linear operators so that the linearized equation can be solved with only one of the linear operators. Such an equation can sometime be easier to solve than the original linearized equation.

The same also applies to GLIM-II as well as to GLIM-I, but not to GLIM-III. The reason is that GLIM-III is not flexible to the extent that a semilinear (locally) Lipschitzian term can be added to the equation, because then the Taylor estimate in terms of a quadratic function may no longer be valid, which is essential for this method. In this respect, both GLIM-I and -II appear to be more flexible.

1. A SPECIAL CASE OF ELLIPTIC REGULARIZATION OF DEGREE $\bar{k} = 1/2$

Let $\{X_j\}$ with $0 \le j \le p$ be a scale of Hilbert-Sobolev spaces with increasing norms such that $i < j$ implies $X_j \subset X_i$ and $\|x\|_j \ge \|x\|_i$, and let $0 < m_1 < m_2 < p$.

(A_0) Denote by $\{S_\theta\}$, $\theta \ge 1$, a one-parameter family of smoothing operators, such that

$$\|(I - S_\theta)x\|_0 \le C\theta^{-m_1}\|x\|_{m_1} \qquad m_1 \ge 0$$

$$\|(I - S_\theta)x\|_{m_1} \le C\theta^{-(m_2-m_1)}\|x\|_{m_2}$$

$$\|S_\theta x\|_p \le C\theta^{-(p-m_2)}\|x\|_{m_2}$$

for some constant $C > 0$, where I is the identity mapping (see Altman, 1984b, c).

We also need the inequality

$$\|x\|_j \le C\|x\|_i^{1-\lambda}\|x\|_p^{\lambda} \tag{1.0}$$

for $j = (1 - \lambda)i + \lambda p$; $0 \le \lambda \le 1$.

Let $W_0 \subset X_s$, with s to be determined, be an open ball with center $x_0 \in X_p$ and radius $r_0 > 0$. Put $V_0 = W_0 \cap X_p$ and let V_s be the $\|\cdot\|_s$-closure of V_0.

Consider the nonlinear operator equation

$$Px \equiv Fx + fx = 0 \tag{1.1}$$

where P is closed on V_s and assumes its values in the same scale of spaces. If the domain of P is $D \subset X_s$, then we put $V_0 = D \cap W_0 \cap X_p$.

We make the following assumptions.

(A_1) F is differentiable; i.e., for each $x \in V_0$, there exists a linear operator $F'(x)$ such that for $h \in X_p$ we have

$$\varepsilon^{-1}\|F(x + \varepsilon h) - Fx - \varepsilon F'(x)h\|_0 \to 0 \qquad \text{as } \varepsilon \to 0+ \tag{1.2}$$

Let $\{x_n\} \subset V_0$ be a Cauchy sequence in X_s and let $\{h_n\}$ be bounded in X_s, then $\varepsilon_n \to 0+$ implies that

$$\varepsilon_m^{-1}\|F(x_n + \varepsilon_n h_n) - Fx_n - \varepsilon_n F'(x_n)h_n\|_0 \to 0 \tag{1.3}$$

There exists a constant $C > 0$ such that

$$\|F'(x)h\|_{m_1} \le C\|h\|_{m_2} \tag{1.4}$$

$$\|F'(x)h\|_0 \le C\|h\|_{m_1} \tag{1.5}$$

$$\|Fx - F'(x)x + fx\|_0 \le C\|x\|_{m_1} \tag{1.6}$$

$$\|fx\|_0 \le C\|x\|_{m_1} \tag{1.7}$$

for all $x \in V_0$.

(A_2) There exists a constant $q_0 > 0$ such that

$$\|f(x + \varepsilon h) - fx\|_0 \le \varepsilon q_0\|h\|_0 \tag{1.8}$$

for all $x \in V_0$.

(A_3) There exists a linear operator A and constants C, $C_0 > 0$, such that

$$\|Az\|_0 \le C\|z\|_{m_2} \tag{1.9}$$

$$\|z\|_{m_2} \le C_0\|Az\|_0 \tag{1.10}$$

$$\|Az\|_0^2 \le (F'(x)z, Az)_0 \qquad x \in V_0 \tag{1.11}$$

For $x \subset V_0$ and η with $0 < |\eta| < 1$, the linearized equation

$$\eta Az + F'(x)z + Fx - F'(x)x + fx = 0 \tag{1.12}$$

has a solution $\bar{z}$.

1.1 LEMMA For $x \in V_0$, if h is a solution of the equation

$$F'(x)h + g = 0 \tag{1.13}$$

then

$$\|h\|_{m_2} \le C_0\|g\|_0 \tag{1.14}$$

with C_0 as in (1.10).

Proof: It results from (1.10), (1.11) that

$$\|h\|_{m_2} \le C_0\|F'(x)h\|_0 \quad \blacksquare$$

To simplify the notation let $0 < \eta < 1$.

1.2 LEMMA If $\bar{z}$ is a solution of the linearized equation (1.12), then

$$\|\bar{z}\|_{m_2} \le C\eta^{-1/2}\|Fx - F'(x)x + fx\|_0 \tag{1.15}$$

for some constant $C > 0$.

Proof: We have, by (1.11),

$$(Az, \eta A\bar{z} + F'(x)\bar{z})_0 \ge \eta\|A\bar{z}\|_0^2 + \|A\bar{z}\|_0^2$$

and

$$(\eta A\bar{z} + F'(x)\bar{z}, A\bar{z})_0 \le 2^{-1}\|\eta A\bar{z} + F'(x)\bar{z}\|_0^2 + 2^{-1}\|A\bar{z}\|_0^2$$

Hence, by (1.10),

$$2\eta C^{-1}\|\bar{z}\|_{m_2}^2 \le 2\eta\|A\bar{z}\|_0^2 + \|A\bar{z}\|_0^2 \le \|\eta A\bar{z} + F'(x)\bar{z}\|_0^2$$

and (1.15) follows by (1.12). $\blacksquare$

1.2a LEMMA Suppose that (1.10) is replaced by

$$\|z\|_{m_2} \le C(\|Az\|_0 + \|z\|_0) \tag{1.10a}$$

and $\bar{z}$ is a solution of the linearized equation

$$\eta(Az + z) + F'(x)z + Fx - F'(x)x + fx = 0 \tag{1.12a}$$

If in addition,

$$(z,Az)_0 \geq \|z\|_0^2 \tag{1.16}$$

then the estimate (1.15) holds.

Proof: We have, by (1.16) and (1.11),

$$(\eta(A\bar{z} + \bar{z}) + F'(x)\bar{z},\ A\bar{z})_0 \geq \eta(\|A\bar{z}\|_0^2 + \|\bar{z}\|_0^2) + \|A\bar{z}\|_0^2$$

and

$$(\eta(A\bar{z} + \bar{z}) + F'(x)\bar{z},\ A\bar{z})_0 \leq 2^{-1}\|\eta(A\bar{z} + \bar{z}) + F'(x)\bar{z}\|_0^2 + 2^{-1}\|A\bar{z}\|_0^2$$

But, by (1.10a), we have

$$\|A\bar{z}\|_0^2 + \|\bar{z}\|_0^2 \geq 2^{-1}(\|A\bar{z}\|_0 + \|\bar{z}\|_0)^2 \geq 2^{-1}C^{-1}\|z\|_{m_2}^2$$

and (1.15) results from the above. ∎

For $x \in V_0$, let $\bar{z}$ be a solution of equation (1.12). Then the following estimates hold.

1.3 LEMMA If $\|x\|_p < K$, $K > 1$; $0 < \eta < 1$, then

$$\|F'(x)(I - S_\theta)\bar{z}\|_0 \leq M_1\theta^{-(m_2-m_1)}\eta^{-1/2}K^{\bar{\nu}} \tag{1.17}$$

$$\|z\|_p \equiv \|S_\theta\bar{z}\|_p \leq M_2\theta^{p-m_2}\eta^{-1/2}K^{\bar{\nu}} \tag{1.18}$$

for some $M_1, M_2 > 0$ and $z = S_\theta\bar{z}$, $\bar{\nu} = (m_1 - \rho)/(p - \rho)$ with ρ to be determined (see Lemma 1.6, Chapter 2).

Proof: We get from (A_0) and (1.5),

$$\|F'(x)(I - S_\theta)\bar{z}\|_0 \leq C\|(I - S_\theta)\bar{z}\|_{m_1} \leq C\theta^{-(m_2-m_1)}\|\bar{z}\|_{m_2}$$

By (1.15) and (1.6),

$$\|\bar{z}\|_{m_2} \leq C_1\eta^{-1/2}\|x\|_{m_1}$$

Since $\|x\|_{m_1} \le C\|x\|_s^{1-\bar{\nu}}\|x\|_p^{\bar{\nu}} \le Cr_0^{1-\bar{\nu}}K^{\bar{\nu}}$, and $x \in V_0$, we get

$$\|\bar{z}\|_{m_2} \le C_2\eta^{-1/2}K^{\bar{\nu}} \tag{1.19}$$

and (1.17) follows from (1.19). Since, by (A_0), $\|S_\theta\bar{z}\|_p \le C\theta^{p-m_2}\|z\|_{m_2}$, relation (1.18) results from (1.19). ■

Consider the linearized equation

$$F'(x)z + Fx - F'(x)x + fx = 0 \tag{1.20}$$

and put $z = S_\theta\bar{z}$, where $\bar{z}$ is a solution of the equation (1.12) satisfying (1.19).

1.4 LEMMA The following estimate holds:

$$\|F'(x)z + Fx - F'(x)x + fx\|_0 \le M_3(\theta^{-m} + \eta)\eta^{-1/2}K^{\bar{\nu}} \tag{1.21}$$

for $x \in V_0$ with $\|x\|_p < K$, where $m = m_2 - m_1$.

Proof: We have, by (1.12), (1.17), and (1.9),

$$\begin{aligned}
&\|F'(x)z + Fx - F'(x)x + fx\|_0 \\
&\le \|\eta A\bar{z} + F'(x)\bar{z} + Fx - F'(x)x + fx\|_0 \\
&\quad + \|F'(x)(I - S_\theta)\bar{z}\|_0 + \eta\|A\bar{z}\|_0 \\
&\le M_1\theta^{-m}\eta^{-1/2}K^{\bar{\nu}} + C_3\eta K^{\bar{\nu}}
\end{aligned}$$

and (1.21) follows. ■

Now suppose there exist ν,μ with

$$\bar{\nu} < \nu < 1 \quad \text{and} \quad \mu(1 - \nu) - \nu > 0 \tag{1.22}$$

such that for $Q > 1$, $K > 1$, one can find $0 < \eta < 1$ and $\Theta > 1$ to satisfy

$$\Theta^{p-m_2}\eta^{-\bar{k}}K^{\bar{\nu}} < QK^{\nu} \tag{1.23}$$

$$(\Theta^{-m} + \eta)\nu^{-\bar{k}}K^{\bar{\nu}} < Q^{-\mu}K^{\nu} \tag{1.24}$$

1.1 DEFINITION Let $\mu > 0$, $0 \leq \nu$, $\sigma \geq 0$ be given numbers. Then the linearized equation (1.20) admits approximate solutions of order (μ,ν,σ) if there exists a constant $M > 0$ with the following property. For every $x \in V_0$, $K > 1$ and $Q > 1$ if $\|x\|_p < K$, then there exist a residual (error) vector y and a vector z such that

$$\|z\|_p \leq MQK^{\nu} \tag{1.25}$$

$$\|y\|_0 \leq MQ^{-\mu}K^{\sigma} \tag{1.26}$$

and

$$F'(x)z + Fx - F'(x)x + fx + y = 0 \tag{1.27}$$

1.5 LEMMA The linearized equation (1.20) admits approximate solutions of order (μ,ν,σ) with $\sigma = \nu$.

Proof: The proof follows from (1.24) and (1.18) with η as in (1.23), where ν will be determined later (see Lemma 1.8). ■

Thus we have shown that if $\bar{z}$ is a solution of equation (1.12), then $z = S_\theta \bar{z}$ with appropriate choice of Θ is an approximate solution of order (μ,ν,σ) with $\sigma = \nu$ of the linearized equation (1.20); that is, there exists a constant M and a vector z that satisfy relations (1.25) to (1.27).

Now put $z = x + h$. Then it follows from (1.27) that h is a solution of the equation

$$F'(x)h + Px + y = 0 \tag{1.28}$$

and we get from (1.25), (1.26), and (1.14),

$$\|h\|_0 \leq C_0(\|Px\|_0 + MQ^{-\mu}K^{\sigma}) \tag{1.29}$$

$$\|h\|_p \leq \|x\|_p + MQK^{\nu} \tag{1.30}$$

where $\sigma = \nu$.

1.6 LEMMA Let $x \in V_0$ with $\|x\|_p < K$, $K > 1$, and let

$$q_0C_0 < q < 1 \quad \text{and} \quad \bar{q} = (q - q_0C_0)/2 < q \tag{1.31}$$

If h is a solution of equation (1.28), then there exists $0 < \varepsilon \leq 1$ such that

$$\|P(x + \varepsilon h) - (1 - \varepsilon)Px\|_0 \leq \varepsilon q\|Px\|_0 \tag{1.32}$$

and

$$\|h\|_0 \leq C_0(1 + \bar{q})\|Px\|_0 \tag{1.33}$$

Proof: For the proof see Lemma 1.1, Chapter 1. ∎

1.7 LEMMA For $x_n \in V_0$ let

$$\|x_n\|_p < A \exp(\alpha(1 - q)t_n) = K_n \tag{1.34}$$

where $t_n > 0$, and let α, A, $\bar{q}$, and q be such that

$$\begin{gathered} \max([\mu(1 - \nu) - \nu]^{-1}, (1 - q)^{-1}) < \alpha \\ \mu(1 - \nu) - \nu > 0 \end{gathered} \tag{1.35}$$

with μ,ν as in Lemma 1.8, and

$$M^{1+1/\mu}(\bar{q}p_0)^{-1/\mu} < A^{1-\nu-\nu/\mu}[\alpha(1 - q) - 1] \qquad \alpha > 1 \tag{1.36}$$

where $p_0 = \|Px\|_0$, and q,q_0 satisfy (1.31). Then there exists a number $Q = Q_n$ such that

$$\|x_n + \varepsilon h_n\|_p \quad A \exp(\alpha(1 - q)(t_n + \varepsilon)) \tag{1.37}$$

for all $0 < \varepsilon \leq 1$, where h_n is a solution of equation (1.28) with $x = x_n$, $y = y_n$, provided that relations (1.29) and (1.30) hold with $x = x_n$, $h = h_n$, $K = K_n$, and $Q = Q_n$.

Proof: For the proof see Lemma 1.2, Chapter 1. ∎

1.8 LEMMA There exist ν with $\bar{\nu} < \nu < 1$ and μ which satisfies (1.22) and such that for $K > 1$ and $Q > 1$ one can find $0 < \eta < 1$ and $\Theta > 1$ which satisfy (1.23) and (1.24) provided

$$0 < \bar{k} < 1 \tag{1.38}$$

if

$$\bar{\nu} = \frac{\bar{p} - \rho}{p - \rho} \tag{1.39}$$

with ρ as in Lemma 1.6, Chapter 2, where

$$0 < \rho < \bar{p} \tag{1.40}$$

Proof: For the proof see Lemma 1.6, Chapter 2. ∎

In the same way as in Chapter 2, we construct the following method of contractor directions,

$$x_{n+1} = x_n + \varepsilon_n h_n \qquad t_{n+1} = t_n + \varepsilon_n \qquad t_0 = 0 \tag{1.41}$$

where h_n is a solution of the equation (1.28) with $x = x_n$, $y = y_n$, and induction assumptions

$$\|x_n\|_p < A \exp(\alpha(1 - q)t_n) = K_n \tag{1.42}$$

$$\|Px_n\|_0 \le \|Px_0\|_0 \exp(-(1 - q)t_n) \tag{1.43}$$

where α, A, and q are subject to (1.35), (1.36) with ν, $0 < s < m_1$, satisfying the assumptions of Lemma 1.8 (see Chapter 2 for more details).

The following estimate holds (see Chapter 2).

$$\sum_{n=0}^{\infty} \varepsilon_n \|h_n\|_s \le CN[\lambda(1 - q)]^{-1} \exp(\lambda(1 - 1)) < r_0 \tag{1.44}$$

where $N = (2Cp_0)^{1-\bar{s}}[\alpha(1 - q)A]^{\bar{s}}$ provided

$$\lambda = 1 - \bar{s}(1 + \alpha) > 0 \qquad \text{or} \qquad \alpha < \frac{1 - \bar{s}}{\bar{s}} \qquad \bar{s} = \frac{s}{p}$$

and the choice of $0 < \varepsilon_n \le 1$ is the same as in Chapter 2. It is clear that the second inequality (1.44) is satisfied if $p_0 = \|Px_0\|_0$ is sufficiently small.

1.1 THEOREM In addition to assumptions (A_0) to (A_3), suppose that condition (1.31) is satisfied and $\|Px_0\|_0$ is so small that (1.44) holds. Then equation (1.1) has a solution x such that $\|x_n - x\|_s \to 0$ as $n \to \infty$, where $\{x_n\} \subset V_0$ is determined by (1.41) and $\|x - x_0\|_s < r_0$.

Proof: The proof is the same as that of Theorem 1.1, Chapter 1. ∎

1.1 REMARK Suppose that

$$Fx = F_1x + F_2x \qquad \text{for } x \in V_0$$

Then Theorem 1.1 holds true if the assumptions for $F'(x)$ are satisfied by $F_1'(x)$ and the estimate for $f(x + \varepsilon h) - fx$ is also valid for F_2x. Then evidently instead of the linearized equation (1.12) we consider

$$\eta Az + F_1'(x)z + Fx - F_1'(x)x + \bar{f}x = 0 \qquad \bar{f}x = F_2x + fx$$

and instead of (1.28) we get

$$F_1'(x)h + Px + y = 0$$

Then we obtain the following identity:

$$\begin{aligned} P(x + \varepsilon h) - (1 - \varepsilon)Px = {} & [F(x + \varepsilon h) - Fx - \varepsilon F'(x)h] \\ & + [f(x + \varepsilon h) - fx] + \varepsilon F_2'(x)h \\ & + \varepsilon[F_1'(x)h + Px + y] - \varepsilon y \end{aligned}$$

1.2 REMARK It follows from (1.10) and (1.11) that

$$\|z\|_{m_2} \leq C\|\eta Az + F(x)z\|_0 \leq C\eta^{-\bar{k}}\|\eta Az + F(x)z\|_0$$

holds for arbitrary $\bar{k} > 0$ which can be chosen so as to satisfy (1.38) (see Lemma 1.8). Therefore the last inequality has an advantage over (1.15).

2. NONLINEAR PARTIAL DIFFERENTIAL EQUATIONS

2.1 EXAMPLE In order to illustrate that the hypotheses of Section 1 are verifiable, let us consider the following Dirichlet problem,

$$\begin{aligned} cAu + F(x,u,Du,D^\beta u) + f(x,u) &= 0 \qquad \text{on } \Omega \\ u &= 0 \qquad \text{on } \partial\Omega \end{aligned} \tag{2.1}$$

where Ω is some open and bounded domain in R^N, $\partial\Omega$ is the boundary of Ω, Du stands for the gradient of u, and $D^k u$ denote the partial derivatives with multi-index β and $|\beta| = \beta_1 + \cdots + \beta_N \le n$. The space X_0 is defined as the completion of C^∞ or C_0^∞ in the norm

$$(u,u)_0 = \|u\|_0^2 = \|u\| = \int_\Omega |u(x)|^2 \, dx$$

where $x = (x_1, \ldots, x_N)$, $dx = dx_1 dx_2 \cdots dx_N$ is the volume element. The following Sobolev norms are defined for X_j $(j = 1, 2, \ldots)$:

$$\|u\|_j^2 = \sum_{|\beta|=0}^{j} \|D^\beta u\|^2$$

We assume that the linear operator A in (2.1) is an elliptic operator of even order m_2 and satisfies the condition

$$\|v\|_{m_2} \le C_0 \|Av\|_0$$

The abstract derivative operator for

$$cAu + F(x,u,Du,D^\beta u) \equiv cAu + Fu$$

is given by the formula

$$cAv + F'(u)v = cAv + a_0(x,u,Du,\ldots,D^\alpha u,\ldots)v + \sum_{1\le|\beta|\le n} a_\beta(x,u,Du,D^\beta u) D^\beta v \tag{2.2}$$

where $a_0 = \partial F/\partial u$, and similarly a_β $(|\beta| \le n)$, are the partial derivatives of F corresponding to the partial derivatives of $v = v(x_1, \ldots, x_N)$. Using Sobolev's inequality one can estimate $|u|$ and $|D^\beta u|$ in terms of $\|u\|_s$, $n + N/2 < s$. Since $\|u\|_s < r_0$, by virtue of Theorem 1.1, it follows that the coefficients a_0 and a_β in (2.2) are bounded by constants $M_0(r_0)$, $M_\beta(r_0)$, provided that F has sufficiently many continuous derivatives. In order to obtain the estimate [see (1.11)]

$$\int_\Omega (cAv + F'(u)v)Avdx \ge \|Av\|_0^2 \tag{2.3}$$

we replace $|v(Av)|$ by $2^{-1}(v^2 + (Av)^2)$ and $|D^\beta v||Av|$ by

$2^{-1}(|D^\beta v|^2 + |Av|^2)$ in the corresponding terms of $(F'(u)v,Av)_0$ obtained by means of (2.2). In this way we obtain

$$(cAv + F'(u)v,\ Av)_0 \geq (c - d)\|Av\|_0^2$$

for some $d > 0$, and (2.3) follows if $c - d \geq 1$. The remaining hypotheses of this section can easily be verified, and Lemma 1.2a is also applicable.

2.2 EXAMPLE Instead of equation (2.1), one can consider the following one,

$$\begin{aligned} cAu + \lambda Bu + F(x,u,Du,D^\beta u) &= 0 \quad \text{on } \Omega \\ u &= 0 \quad \text{on } \partial\Omega \end{aligned} \tag{2.4}$$

where A and F are the same as in (2.1) and λ is a sufficiently small number. Then equation (2.4) can be reduced to the equation (2.1) by putting in (2.1) $f(x,u) = Bu$, and $q_0 = \lambda$ in (1.8).

For existence of solutions of the equations of the type (1.12), consult Agmon et al. (1959; 1964).

3. SYSTEMS OF NONLINEAR PARTIAL DIFFERENTIAL EQUATIONS

Let us consider the following Dirichlet problem,

$$H_k(x,u,Du,D^\beta u) = 0 \quad \text{on } \Omega \quad \text{and} \quad u = 0 \quad \text{on } \partial\Omega \tag{3.1}$$

where $k = 1, \ldots, m$, $x = (x_1,\ldots,x_N)$, $u = (u_1,\ldots,u_m)$, $u_k = u_k(x)$ $(k = 1,\ldots,m)$, and $|\beta| \leq n$. Thus the equation (2.1) is replaced by the system (3.1). As in Example 2.1, we consider the following class of systems for which the estimate (1.11) can easily be verified under suitable conditions:

$$H_k(x,u,Du,D^\beta u) = cA_k u_k + F_k(x,u,Du,D^\beta u) + f_k(x,u) = 0 \tag{3.2}$$

where $c > 0$ is a constant and Du_k is the gradient of u_k, $D^\beta u$ being higher order partial derivatives.

The following norms, $\|\cdot\|_j$, are defined.

$$(u,u)_j = \|u\|_j^2 = \sum_{i=1}^{m} \|u_i\|_j^2 \qquad \text{with} \qquad \|u_i\|_j^2 = \sum_{|\beta|=0}^{j} \|D^\beta u_i\|^2$$

$$(u_i,u_i)_0 = \|u_i\|_0^2 = \|u_i\|^2 = \int_\Omega |u_i(x)|^2\, dx \qquad |\beta| = \beta_1 + \cdots + \beta_N$$

for $j = 0, s, m_1, m_2, r$ with $0 < s < m_1, 3m_1 < m_2 < r$ (s to be determined).

The Hilbert-Sobolev spaces $X_j(\|\cdot\|_j)$ are obviously such that $X_j \subset X_{j'}$ for $j \geq j'$ and $\|\cdot\|_j \geq \|\cdot\|_{j'}$.

Let $W_0 \subset X_s$, with s to be determined, be an open ball with center $u^{(0)}$ in X_r and radius $r_0 > 0$. Put $V_0 = W_0 \cap X_r$ and let V_s be the closure in X_s of V_0

To simplify the notation let us assume that $|\beta| = 1$. The functions $F_k(x,u,p)$ admit sufficiently many derivatives in some region $|u| + |p| \leq c(r_0)$, where p has components p_{ki} corresponding to $\partial u_k/\partial x_i$ and the matrices $a^{(i)}$ are defined by the entries

$$a_{kj}^{(i)}(x,u,p) = \frac{\partial F_k}{\partial p_{ji}}$$

for $k,j = 1, \ldots, m$; $i = 1, \ldots, N$, and the matrix b is given by

$$b_{kj}(x,u,p) = \frac{\partial F_k}{\partial u_j}$$

Put $H(x,u,p) = (H_1(x,u,u_x),\ldots,H_m(x,u,p))$,

$$Fu = (F_1(x,u,p),\ldots,F_m(x,u,p))$$

$$fu = (f_1(x,u),\ldots,f_m(x,u))$$

$$Pu = H(\cdot,u,p)$$

The abstract derivative of F is given by the formula

$$F'(u)v = \sum_{i=1}^{N} a^{(i)} \frac{\partial}{\partial x_i} v + bv \tag{3.3}$$

Now let A_k be the same type linear differential operators as A in the equation (2.1), and we assume that the Dirichlet problem for the modified linearized equation [see (1.12)] for (3.2) has a solution provided that F, F', f, and u are replaced by the same expressions defined above and $Av = (A_1v_1,\dots,A_mv_m)$. Also assume that A_k satisfies condition (1.10), i.e.,

$$\|v_k\|_{m_2} \le C_0\|A_kv_k\|_0 \tag{3.4}$$

for $k = 1, \dots, m$, where $v = (v_1,\dots,v_m)$. Now put

$$G_k = A_kv_k \qquad \text{and} \qquad G = (G_1,\dots,G_m) \tag{3.5}$$

We have

$$\int_\Omega G_kAv_k\,dx = \int_\Omega |A_kv_k|^2\,dx = \|A_kv_k\|_0^2$$

Hence, we obtain

$$\int_\Omega G\cdot(Av)\,dx = \sum_{k=1}^m \|A_kv_k\|_0^2 = \|Av\|_0^2 \tag{3.6}$$

Next we have by (3.3)

$$\begin{aligned}(F'(u)v,Av)_0 &= \left(\sum_{i=1}^N a^{(i)}\frac{\partial}{\partial x_i}v,Av\right)_0 + (bv,v)_0\\
&= \int_\Omega \sum_{i=1}^N\sum_{k=1}^m\left(\sum_{j=1}^m a_{kj}^{(i)}\frac{\partial}{\partial x_i}v_j\right)(A_kv_k)\,dx\\
&\quad + \int_\Omega \sum_{k=1}^m\left(\sum_{j=1}^m b_{kj}v_j\right)(A_kv_k)\,dx\\
&= \int_\Omega \sum_{i=1}^N\sum_{k=1}^m\left(\sum_{j=1}^m a_{kj}^{(i)}\frac{\partial}{\partial x_i}v_j\right)(A_kv_k)\,dx\\
&\quad + \int_\Omega \sum_{k=1}^m\left(\sum_{j=1}^m b_{kj}v_j\right)(A_kv_k)\,dx\end{aligned}$$

But

$$\left(\frac{\partial}{\partial x_i}v_j\right)(Av_k) \le 2^{-1}\left[\left(\frac{\partial}{x_i}v_j\right)^2 + (A_kv_k)^2\right]$$

and

$$(v_j)(A_k v_k) \le 2^{-1}[(v_j)^2 + (A_k v_k)^2]$$

and assuming the following estimates,

$$\sum_{i=1}^{n} |a_{kj}^{(i)}(x,u,Du)| \le M(r_0) \quad \text{and} \quad |b_{kj}(x,u,Du)| \le M_0(r_0) \tag{3.7}$$

which follow from the Sobolev inequality, provided that $\|u\|_s < r_0$ for some s, $1 + n/2 < s$ to be determined, we get from (3.4) with $C_0 \ge 1$

$$(F'(u)v,Av)_0 \le C_0 2^{-1} M(r_0) \sum_{k=1}^{m} \sum_{j=1}^{m} (\|v_j\|_1^2 + \|Av_k\|_0^2)$$

$$+ C_0 2^{-1} M_0(r_0) \sum_{k=1}^{m} \sum_{j=1}^{m} (\|v_j\|_0^2 + \|Av_k\|_0^2)$$

Hence, by (3.4), we obtain

$$(F'(u)v,Av)_0 \le C_0 m(M(r_0) + M_0(r_0))\|Av\|_0^2 \tag{3.8}$$

Relations (3.6) and (3.8) imply (1.11) with F' replaced by G' + F' given above, provided that

$$c - C_0 m(M(r_0) + M_0(r_0)) \ge 1 \tag{3.9}$$

One can assume that F_k (k = 1,...,m) has sufficiently many derivatives so that the assumptions (A_1) and (A_2) can be verified. Also $m_1 < m_2 < r$ should satisfy condition (1.40) with p replaced by r. Thus we have shown that the assumptions of Section 1 are verifiable.

Let us mention that Lemma 1.2a can also be applied here.

3.1 REMARK The same method is also applicable to the Dirichlet problem for systems of higher order of the form

$$A_k u_k + F_k(x,u,Du,\ldots,D^\beta u,\ldots) + f_k(x,u) = 0 \tag{3.10}$$

where k = 1, ..., m. In this case, s should satisfy the condition

$n + N/2 < s$, where n is highest order of partial derivatives of u in (3.10).

3.2 REMARK In the same way as in Example 2.2, one can replace A_k in (3.10) by $A_k + \lambda B_k$, where B_k is semilinear. Then λB_k becomes a term like f_k in (3.10) and as such can be added to f_k.

3.3 REMARK Remark 1.2 also applies to Examples 2.1 and 2.2 and to problem (3.1), Section 3, as well.

4. PERIODIC SOLUTIONS OF NONLINEAR HYPERBOLIC PARTIAL DIFFERENTIAL EQUATIONS

Rabinowitz (1969) has studied the nonlinear dissipative wave equation

$$Pu \equiv u_{tt} - u_{xx} + \alpha u_t + \varepsilon F(x,t,u,u_x,u_t,u_{xx},u_{tx},u_{tt}) = 0 \qquad (4.1)$$

with F being periodic in t, and x-interval normalized to be 1. He proved the existence of a classical solution to (5.1) satisfying the periodicity and the boundary conditions

$$u(x, t + \tau) = u(x,t), \; u(0,t) = u(1,t) = 0 \qquad (4.2)$$

provided that F is sufficiently smooth and ε is sufficiently small. As a by-product of his hyperbolic existence theorem he obtained an analogous parabolic result, for

$$u_t - ku_{xx} + \varepsilon F(x,t,u,u_x u_t,u_{xx},u_{xt},u_{tt}) = 0 \qquad (4.3)$$

satisfying (5.2).

Using Rabinowitz's notation, let C_0^∞ be the subspace of C^∞ of τ-periodic in t functions which vanish near x = 0, and x = 1 with $0 \le x \le 1$. Let H_0 be the completion of C^∞ in the norm

$$|u|_0^2 = |u|^2 = \int_0^1 \int_0^\tau |u(x,t)| \; dx \; dt$$

and $\overset{0}{H}_j$ the completion in the norm

$$|u|_j^2 = \sum_{|\beta|=0}^{1} |D^\beta u|^2$$

where $\beta = (\beta_1, \beta_2)$, $|\beta| = \beta_1 + \beta_2$, $D^\beta = \partial^{\beta_1+\beta_2}/\partial x^{\beta_1}\partial t^{\beta_2}$, and let H_j be the completion of C_0^∞ with respect to $|\cdot|_j$.

Let C_j denote the space of j times continuously differentiable functions in x and t which are τ-periodic in t, $0 \le x \le 1$. The norm in C_0 is

$$\|u\|_0 = \|u\| = \sup_{x,t} |u(x,t)|$$

and the norm in C_j is

$$\|u\|_j = \sum_{|\beta|=0}^{j} \|D^\beta u\|$$

Let

$$H = \{u \in \overset{0}{H}_1 \,|\, u \in H_{k+1},\ u_t \in H_{k+1}\} \quad \text{with} \quad |u|_H = |u|_{k+1} + |u_t|_{k+1}$$

$$\tilde{H} = \{u \in H_1 \,|\, u \in H_{k+1},\ u_t \in H_{k-1}\} \quad \text{with} \quad |u|_{\tilde{H}} = |u|_{k-1} + |u_t|_{k-1}$$

and

$$\hat{H} = \{u \in \overset{0}{H}_1 \,|\, u \in H_{k+1},\ u_{t^{2j+1}} \in H_{k+1}\}$$

$$|u|_{\hat{H}} = |u|_{k+1} + |u_{t^{2j+1}}|_{k+1}$$

where j is to be defined in terms of k. Let

$$D = H \cap \{u \in H_0 \,|\, \|u\|_2, \|u_t\|_2 < 1\}$$

Rabinowitz proved that $P\colon\ D \to \tilde{H}$. Let

$$P'(u)v = \lim_{\varepsilon \to 0} [P(u + \varepsilon v) - Pu] \qquad u \in D,\ v \in H$$

Then one obtains

$$P'(u)v = u_{tt} - v_{xx} + \alpha v_t + \varepsilon\{F_u(x,t,u,\ldots)v + F_{u_t}(\cdot)v_t$$
$$+ F_{u_x}(\cdot)v_x + F_{u_{xx}}(\cdot)v_{xx} + F_{u_{xt}}(\cdot)v_{xt} + F_{u_{tt}}(\cdot)v_{tt}\}$$
$$\equiv v_{tt} - v_{xx} + \alpha v_t + \varepsilon A(u)v$$

where

$$A(u)v \equiv a_0(x,t,u,\ldots)v + a_{10}(x,t,u,\ldots)v_x + a_{01}(\cdots)v_t$$
$$+ a_{20}(\cdots)v_{xx} + a_{11}(\cdots)v_{xt} + a_{02}(\cdots)v_{tt}$$

Put

$$|A|_i = |a_0|_i + |a_{10}|_i + \cdots + |a_{02}|_i$$
$$\|A\|_i = \|a_0\|_i + \|a_{10}\|_i + \cdots + \|a_{02}\|_i$$

Using highly technical arguments Rabinowitz (1969) found a solution to the following linearized equation,

$$Lv = (-1)^{j+1}\eta^{2j+1}\Delta v_{t^{2j+1}} + P'(u)v = g \tag{4.4}$$

$\Delta = \partial^2/\partial t^2 + \partial^2/\partial x^2$, $0 < \eta < 1$ (parameter), with j chosen in terms of k, and he obtained the following estimate for the solution,

$$\eta^{2j+1}|v_{t^{2j+1}}|_{k+1} + |v|_{k+1} \le \text{const}\left[|g|_{k=1} + \varepsilon|A|_{k-1} + |g_t|_{k-1} + \varepsilon\left|\frac{\partial}{\partial t}A\right|_{k-1}\right] \tag{4.5}$$

$$\|v\|_2 \le 1 \qquad 2j + 1 = k - 1$$

provided that $|g|_2 + |g_{ttt}|$ and $\varepsilon(\|A\|_2 + \|A_{ttt}\|)$ are sufficiently small. This is the crucial result of Rabinowitz's work, which is essential for the application of the well-known Moser (1966) theorem in order to prove the existence theorem for (4.1) and (4.2). He also makes use of the Moser type inequality

$$|F(x,t,u,\ldots,u_{tt})|_{k-1} \le c(|u|_{k+1} + 1) \tag{4.6}$$

for "composition of functions," which is an extension of the inequality

$$|f(x,t,u)|_i \le c(|u|_i + 1) \tag{4.7}$$

Rabinowitz has also proved that

$$|P'(u)v|_0 \ge \text{const}\,|v|_1 \tag{4.8}$$

Now let us consider equation (4.1) with u satisfying (4.2). Suppose that F in (4.1) satisfies all the assumptions made by Rabinowitz in order to solve problems (4.1) and (4.2), which include

$$|Q(u,v)|_0 \le \varepsilon\ \text{const}\|v\|_2 |v|_2$$

where $Q(u,v) = \varepsilon F(x,t,u + v,\ldots) - \varepsilon F(t,x,u,\ldots) - \varepsilon A(u)v$. Rabinowitz's result can be obtained by making use of Theorems 3.1 or 3.2, Chapter 2.

In fact, the linearized equation to be solved is the same as (4.4) with a change of g only, i.e.,

$$g = -\varepsilon[F(u) - F'(u)u]$$

As far as the implications of Lemma 1.6 or 3.4, Chapter 2, are concerned, the inequality (4.5) can be replaced by the following one having the same effect,

$$|v|_{k+1} \le c\eta^{-1/(k-1)} |g|_{k-1}$$

where $2j + 1 = k - 1$ in (4.4) and η^{k-1} in (4.4) is replaced by a new $\bar{\eta} = \eta^{k-1}$ and writing η instead of $\bar{\eta}$. It follows from the above that the hypotheses of Chapter 2 are verifiable. Rabinowitz has proved that the hypotheses of his existence theorem are satisfied for $k = 16$.

4.1 REMARK By the same argument the inequality (4.5), in which $|v|_{k+1}$ is dominated by the right-hand side, can be replaced by the following

$$|v|_{k+1} \le C|g|_{k-1} \le C\eta^{-\bar{k}} |g|_{k-1}$$

where $0 < \eta < 1$ and $\bar{k} > 0$ is arbitrary. It follows then that each of the three methods of Chapter 2 is applicable.

5. LINEAR EQUATIONS WITH SMALL NONLINEAR PERTURBATIONS

Petzeltova (1983) proved an existence theorem for the equation

$$Lu = \varepsilon F(u) \tag{5.1}$$

where L is linear, F is nonlinear, and $\varepsilon > 0$ is sufficiently small. The proof given by Petzeltova is based on Moser's theorem (1966).

Let $H_0 \supset \hat{H}_0 \supset H_1 \supset \hat{H}_1 \supset \cdots \supset H_k \supset \hat{H}_k \cdots$ be Hilbert spaces with norms such that

$$|u|_0 \le \|u\|_0 \le |i|_1 \le \cdots \le |u|_k \le \|u\|_k \cdots$$

where $|\cdot|_i$, $\|\cdot\|_i$ are the norms in H_i, $\hat{H}_i$, respectively, and

$$|u|_{k+i+1} \le c|u|^{1-s}|u|^s_{k+i+2} \qquad 0 < s < 1$$

$$\|u\|_\ell \le c\|u\|_0^{1-\sigma}\|u\|_k^\sigma \qquad 0 < \ell < k,\ 0 < \sigma < 1$$

all constants being denoted by c.

Let L be a linear operator continuously mapping $H_{\ell+i} \to H_\ell$, $\hat{H}_{\ell+i} \to \hat{H}_\ell$ for $0 \le \ell \le k$ and some $i > 0$, and let F be a nonlinear differentiable operator with the same properties.

Assumptions: Let $u_0 \in \hat{H}_{k+i}$,

$$D = \{u \in \hat{H}_{k+i},\ \|u - u_0\|_\ell \le R\} \qquad T(u) = Lu - \varepsilon F(u)$$

Then there exists a constant $M > 0$ such that

(i) $\|F(u)\|_k \le MK$ for $u \in \hat{H}_{k+i} \cap D$, $\|u\|_{k+i} \le K$

(ii) $\|T(u) - T(u_0)\|_0 \le$ for $u \in D$

(iii) $\|F(u + v) - F(u) - F'(u)v\|_0 \le M\|v\|_0^{2-\beta}\|v\|^\beta_{k+i}$

for $u \in D$, $v \in \hat{H}_{k+i}$, $0 < \beta < 1$

5.1 THEOREM (Petzeltova) Let the spaces H_j and $\hat{H}_j$ $(0 \le j \le k + i + 2)$ satisfy condition (5.2), and let $\{e_n\}$ be an orthogonal basis in each H_j, e_n being the eigenvectors of the operator L. Suppose that there exists a continuous operator Λ: $\hat{H}_{\ell+i} \to H_\ell$ such that $\Lambda e_n = b_n e_n$, n = 1, 2, ..., and

$$(Lv, \Lambda v)_\ell \geq c|v|^2_{\ell+i}$$

$$(F'(u)v, \Lambda v)_k \leq c|v|^2_{k+i}$$

$$(g, \lambda v)_k \leq c\|g\|_k |v|_{k+i}$$

where $u, v \in \hat{H}_{k+i}$, $\|u\|_{k+i} \leq K$, $0 \leq \ell \leq k + 2$. Then the linearized equation $T'(u)v = g$ admits approximate solutions of degree μ in the sense of Moser, i.e., for every $Q > 1$ there exists $v_Q \in \hat{H}_{j+i}$ such that

$$\|T'(u)v_Q - g\|_0 \leq MKQ^{-\mu}$$

$$\|v_Q\|_{k+i} \leq MKQ$$

$$\|v_Q\|_0 \leq c\|T'(u)v_Q\|_0$$

$$\frac{\mu + 1}{\mu - \lambda} < \min\left(2 - \beta\,\frac{\lambda + (\lambda + 1)(\mu + 1)}{\lambda(\mu + \beta)},\ \lambda\,\frac{1 - \sigma}{\sigma}\right)$$

provided that $\|g\|_0 < K^{-\lambda}$, $\|g\|_k \leq K$, $\|u\|_{k+i} \leq K$, K being sufficiently large. Hence the equation (5.1) has a solution in $\hat{H}_\ell$ for sufficiently small ε.

The proof of Theorem 5.1 is based on the existence of an operator Λ' such that for all $v \in H_{k+i+2}$ and $\eta > 0$ the following conditions are satisfied:

$$|\Lambda' v|_1 \leq c|v|_{k+i}$$

$$|\Lambda' v|_k \geq c|v|_{k+i+2}$$

$$(Lv, \Lambda' v)_k \geq 0 \qquad (\Lambda' v, \Lambda v)_k) \geq 0$$

$(\Lambda + \eta\Lambda')^{-1}$ exists and is continuous from H_k to H_{k+i+2}. The operator $(\eta\Lambda' + T'(u))^{-1}$: $H_k \to H_{k+i+2}$ exists and satisfies the following inequality,

$$|v|_{k+i} + \eta|v|_{k+i+2} \leq c\|\eta\Lambda'v + T'(u)v\|_k \qquad \eta > 0 \tag{5.3}$$

5.1 REMARK The inequality (5.3) yields

$$|v|_k \leq c\|\eta\Lambda'v + T'(u)v\|_k \leq c\eta^{-\bar{k}}\|\eta\Lambda'v + T'(u)v\|_k \tag{5.4}$$

for arbitrary $0 < \eta < 1$ and $\bar{k} \geq 0$. Thus $\bar{k}$ can be chosen so as to satisfy the condition $0 < \bar{k} < 1$. Consequently, if one assumes the existence of smoothing operators, then each of the three methods discussed in Chapter 2 can be applied to prove Theorem 5.1. However, without smoothing operators the stronger inequality (5.3) is needed.

Petzeltova (1983) applied Theorem 5.1 in order to investigate and prove the existence of a periodic solution to the equation

$$Lu = u_{tt} + (-1)^p\partial^{2p}/\partial x^{2p} + \alpha u_t - u = \varepsilon F(u) \tag{5.5}$$

where $F(u) = f(t,x,u,u_t,u_x,u_{tt},u_{tx},u_{xx})$ in case of $p = 1$. If f is sufficiently smooth and ε is sufficiently small, then the problem (5.5), (5.6) has a classical solution under a certain assumption made in Theorem 2 of Petzeltova. The same procedure is applied there for $p = 2, 3, \ldots$.

4

Nonlinear Equations in a Scale of Hilbert-Sobolev Spaces, II

A further development of theory of global linearization iterative methods (GLIM) is presented in this chapter. Also, a method of elliptic regularization of degree $\bar{k} = 1/2$ is proposed which is based on the assumption that the linear abstract derivative operator is positive definite. We also use here elliptic regularization combined with smoothing operators. Although the method under consideration still works in case of GLIM-II, it turns out that it may not be applicable to GLIM-III, especially if the nonlinear equation in question contains an extra semilinear, locally Lipschitzian term. Such a term may spoil the positive definiteness, or the Taylor estimate by means of a quadratic function may no longer be valid or both facts may occur. Let us mention that the second fact is also essential in general case of GLIM-III. In this respect both GLIM-I and -II are more flexible.

The linear abstract derivative operator which appears in the linearized equation can be split into a sum of two linear operators in such a way that the linearized equation can be solved with only one of the linear operators. Such an equation can sometimes be easier to solve than the original linearized equation. This feature is characteristic also in general case of both GLIM-I and -II. However, this does not apply to GLIM-III which is not as flexible. A general existence theorem is proved which is related to GLIM-I and an application to the system of first-order nonlinear partial differential equations containing an extra locally Lipschitzian semilinear term is given, i.e., to the system investigated by Moser (1966).

1. A NONLINEAR EXISTENCE THEOREM

Moser (1966) used the following Sobolev spaces X_ρ ($\rho = 0,1,\dots$) of real functions $u(x)$, $x = (x_1,\dots,x_n)$ of period 2π in each $0 \le x_i \le 2\pi$ with inner product

$$(u,v)_\rho = \int u \cdot (-\Delta)^\rho v \, dx$$

where $\Delta = \Sigma_{i=1}^n (\partial/\partial x_i)^2$ is the Laplacean operator, and integration is taken over $0 \le x_i \le 2\pi$ and dx is the volume element $dx_1 \cdots dx_n$. Then $\|u\|_\rho = (u,u)_\rho^{1/2}$ is a seminorm which vanishes for constant functions if $\rho > 0$, but

$$\|u\|_{X_\rho} = (\|u\|_0^2 + \|u\|_\rho^2)^{1/2} \tag{1.0}$$

represents a proper norm. The closure of all C^∞ functions (of period 2π) under this norm is a Hilbert-Sobolev space X_ρ. Using the Fourier expansion

$$v = \Sigma_k v_k e^{ikx} \qquad k = (k_1,\dots,k_n) \qquad k_i\text{-integers}$$

one defines for real $\rho \ge 0$ the norm

$$\|v\|_\rho^2 = 2\pi \, \Sigma_k |k|^{2\rho} |v_k|^2$$

where $|k|^2 = k_1^2 + \cdots + k_n^2$. The closure of trigonometric polynomials in this norm is the space X_ρ. Both norms agree for integers $\rho > 0$. The above norms allow the introduction of smoothing operators. Consider the truncated Fourier series

$$w = \Sigma_{|k| \le N} v_k e^{ikx}$$

with an appropriate integer $N \geq 1$. We get for $m \geq n$,

$$\sum_{|k|>N} (N^{-1}|k|)^{2n}|v_k|^2 \leq \sum_{|k|>N} (N^{-1}|k|)^{2m}|v_k|^2$$
$$\leq \sum_{|k|\geq 0} (N^{-1}|k|)^{2m}|v_k|^2$$

Hence,

$$\|v - w\|_n \leq N^{-(m-n)} \|v\|_m$$

On the other hand, we have for $m \geq n$,

$$\sum_{-N}^{N} (N^{-1}|k|)^{2m}|v_k|^2 \leq \sum_{|k|\leq N} (N^{-1}|k|)^{2n}|v_k|^2$$
$$\leq \sum_{|k|\geq 0} (N^{-1}|k|)^{2n}|v_k|^2$$

Hence,

$$\|w\|_m \leq N^{(m-k)} \|v\|_n$$

Now for arbitrary $\Theta \geq 1$, one can choose an integer N large enough so that

$$N - 1 \leq \Theta \leq N \quad \text{with} \quad N \geq 2$$

and put

$$S_\Theta v = w = \sum_{-(N-1)}^{N-1} v_k e^{ikx}$$

and take $C = C(m,n) = 2^{m-n}$. Then

$$(N - 1)^{-(m-n)} \leq CN^{-(m-n)} \leq C\Theta^{-(m-n)}$$

is satisfied. It follows from the above that the relations

$$\|(I - S_\Theta)v\|_n \leq C\Theta^{-(m-n)} \|v\|_m$$

$$\|S_\Theta v\|_m \leq C\Theta^{m-n}\|v\|_n$$

are fulfilled for arbitrary $m \geq n$ and $\Theta \geq 1$. One can also use the orthogonality of the functions e^{ikx} with respect to all inner products $(u,v)_m$, in order to prove the above norm relations. The following relations can also be verified (see Moser, 1966),

$$\|v\|_j \leq \|v\|_k^{1-\lambda}\|v\|_i^{\lambda}$$

for $j = (1 - \lambda)k + \lambda i$; $0 \leq \lambda \leq 1$. In particular, $\|v\|_\rho \leq \|v\|_0^{1-\rho/r} \times \|v\|_r^{\rho/r}$ for $0 < \rho < r$. We note that

$$X_\rho \supset X_r \quad \text{if } 0 \leq \rho \leq r$$

and it is also clear that

$$\|\cdot\|n \leq \|\cdot\|m$$

for $n \leq m$. For more details see Moser (1966).

Let $W_0 \subset X_s$, with s to be determined, be an open ball with center $u_0 \in X_p$ and radius $r_0 > 0$. Put $V_0 = W_0 \cap X_p$ and let V_s be the closure in X_s of V_0.

Consider the nonlinear operator equation

$$Pu \equiv Fu + fu = 0 \tag{1.1}$$

where P is closed on V_s and assumes its values in X_0. If the domain of P is $D \subset X_s$, then we put $V_0 = D \cap W_0 \cap X_p$.

We make the following assumptions.

(A_1) F is differentiable, i.e., for each $u \in V_0$, there exists a linear operator $F'(u)$ such that for $h \in X_p$

$$\varepsilon^{-1}\|F(u + \varepsilon h) - Fu - \varepsilon F'(u)h\|_0 \to 0 \quad \text{as } \varepsilon \to 0+ \tag{1.2}$$

Let $\{u_n\} \subset V_0$ be a Cauchy sequence in X_s and let $\{h_n\}$ be bounded in X_s; then $\varepsilon_n \to 0+$ implies

$$\varepsilon_n^{-1}\|F(u_n + \varepsilon_n h_n) - Fu_n - \varepsilon_n F'(u_n)h_n\|_0 \to 0 \tag{1.3}$$

There exist constants $C > 0$, m_1 and $m_2 > 2m_1$ such that

$$\|F'(u)h\|_{m_1} \leq C\|h\|_{m_2} \tag{1.4}$$

$$\|F'(u)h\|_0 \le C\|h\|_{m_1} \tag{1.5}$$

$$\|Fu - F'(u)u + fu\|_0 \le C\|u\|_{m_1} \tag{1.6}$$

$$\|fu\|_0 \le C\|u\|_{m_1} \tag{1.7}$$

for all $u \in V_0$.

(A_2) There exists a positive constant $q_0 < 1$ such that

$$\|f(u + \varepsilon h) - fu\|_0 \le \varepsilon q_0\|h\|_0 \tag{1.8}$$

for all $u \in V_0$.

(A_3) The modified linearized equation

$$\eta(-\Delta)^{m_3}v + F'(u)v + Fu - F'(u)u + fu = 0 \tag{1.9}$$

has a solution if $u \in V_0$ and $0 < \eta < 1$, and

$$\begin{aligned} \|v\|_0^2 &\le (F'(u)v,v)_0 \\ \|v\|_{m_1}^2 &\le c(F'(u)v,v)_{m_1} + K_1^2\|v\|_0^2 \end{aligned} \tag{1.10}$$

for all $u \in V_0$, some K_1 and $m_2 - m_1 < m_3 \le m_1$.

1.1 REMARK In order to solve the linear equation

$$L_\eta w = (\eta(-\Delta)^\alpha + L)w = g$$

with the linear operator L, Moser (1966) makes the following assumptions

$$\|v\|_0^2 \le (Lv,v)_0$$

$$\|v\|_m^2 \le c((Lv,v)_m + K_1^2|v|_0^2) \qquad K_1 > 1$$

for $m = r - 1$ and all $v \in X_r$, and

$$\|Lv\|_m \le c\|v\|_r \qquad \text{where } |v|_0 = \sup_x|v|$$

1.1 LEMMA If h is a solution of the equation

$$F'(u)h + g = 0 \tag{1.11}$$

where $u \in V_0$, then

$$\|h\|_0 \le \|g\|_0 \tag{1.12}$$

Proof: It results from (1.10) that

$$\|F'(u)h\|_0 \ge \|h\|_0$$

for all $u \in V_0$, $h \in V_{m_2}$. Hence, relation (1.12) follows. ■

1.2 LEMMA If v is a solution of the equation

$$L_\eta v = \eta(-\Delta)^{m_3} v + F'(u)v = g \tag{1.13}$$

then with $\bar{p} = m_3 + m_1$, we have

$$\|v\|_{\bar{p}} \le C\eta^{-\bar{k}}\|g\|_{m_1} \qquad \text{with} \qquad \bar{k} = \frac{1}{2} \tag{1.14}$$

Proof: We have by (1.10)

$$\begin{aligned}\eta\|v\|_{\bar{p}}^2 + \|v\|_{m_1}^2 &\le c[(L_\eta v, v) + K_1^2\|v\|_0^2] \\ &\le 2^{-1}(c\|g\|_{m_1}^2 + \|v\|_{m_1}^2) + cK_1^2\|g\|_0^2\end{aligned}$$

which yields (1.14).

It results from Lemma 1.2 and (1.6) that if $\bar{z}$ is a solution of the equation (1.9) then

$$\|\bar{z}\|_{\bar{p}} \le C\eta^{-1/2}\|u\|_{m_2} \tag{1.15}$$

for some constant $C > 0$. ■

1.3 LEMMA If $\|u\|_p < K$, $K > 1$, and $0 < \eta < 1$, then

$$\|F'(u)(I - S_\theta)\bar{z}\|_0 < M_1\Theta^{-(m_2-m_1)}\eta^{-1/2}K^{\bar{\nu}} \tag{1.16}$$

$$\|z\|_p \equiv \|S_\theta \bar{z}\|_p \le M_2 \theta^{p-\bar{p}} \eta^{-1/2} K^{\bar{\nu}} \tag{1.17}$$

for some $M_1, M_2 > 0$ and $z = S_\theta \bar{z}$ and $\bar{\nu} = (\bar{p} - \rho)/(p - \rho)$ with ρ to be determined.

Proof: We get from (1.5),

$$\begin{aligned}\|F'(u)(I - S_\theta)\bar{z}\|_0 &\le C\|(I - S_\theta)\bar{z}\|_{m_1} \\ &\le C\Theta^{-(m_2 - m_1)} \|\bar{z}\|_{m_2}\end{aligned}$$

Since

$$\|u\|_{m_2} \le C\|u\|_\rho^{1-\bar{\nu}} \|u\|_p^{\bar{\nu}} \le C r_0^{1-\bar{\nu}} K^{\bar{\nu}} \tag{1.18}$$

if $u \in V_0$, we get (1.16), by virtue of (1.15). But $\|S_\theta \bar{z}\|_p \le C\Theta^{p-\bar{p}} \|\bar{z}\|_{\bar{p}}$ and (1.17) results from (1.18) and (1.15). ∎

Consider the linearized equation

$$F'(u)z + Fu - F'(u)u + fu = 0 \tag{1.19}$$

and put $z = S_\theta \bar{z}$, where $\bar{z}$ is a solution of equation (1.9) satisfying

$$\|\bar{z}\|_{\bar{p}} \le C\eta^{-1/2} K^{\bar{\nu}} \tag{1.20}$$

for some $C > 0$, by virtue of (1.15) and (1.18).

1.4 LEMMA The following estimate holds

$$\|F'(u)z + Fu - F'(u)u + fu\|_0 \le M(\Theta^{-m} + \eta)\eta^{-1/2} K^{\bar{\nu}} \tag{1.21}$$

for $u \in V_0$ with $\|u\|_p < K$ and $m = m_2 - m_1$.

Proof: We have by (1.16)

$$\begin{aligned}&\|F'(u)z + Fu - F'(u)u + fu\|_0 \\ &\le \|\eta(-\Delta)^{m_3}\bar{z} + F'(u)\bar{z} + Fu - F'(u)u + fu\|_0 \\ &\quad + \|F'(u)(I - S_\theta)\bar{z}\|_0 + \eta\|(-\Delta)^{m_3}\bar{z}\|_0\end{aligned}$$

$$\leq M_1 \Theta^{-m} \,^{-1/2} K^{\bar{\nu}} + \eta \|\bar{z}\|_{\bar{p}}$$

and (1.21) follows by (1.20).

Now suppose there exist ν, μ with $\bar{\nu} < \nu < 1$ and $\mu(1 - \nu) - \nu > 0$ such that for $K > 1$, $Q > 1$, one can find $0 < \eta < 1$ and $\Theta > 1$ to satisfy

$$\Theta^{p-\bar{p}} \eta^{-1/2} K^{\bar{\nu}} < QK^{\nu} \tag{1.22}$$

and

$$(\Theta^{-m} + \eta)\eta^{-1/2} K^{\bar{\nu}} < K^{\nu} Q^{-\mu} \tag{1.23}$$

Then relation (1.21) can be written as

$$\|F'(u)z + Fu - F'(u)u + fu\|_0 \leq 2M_3 Q^{-\mu} K^{\nu} \quad \blacksquare \tag{1.24}$$

1.1 DEFINITION Let $\mu > 0$, $\nu \geq 0$, $\sigma \geq 0$ be given numbers. Then the linearized equation (1.19) admits approximate solutions of order (μ,ν,σ) if there exists a constant $M > 0$ with the following property. For every $u \in V_0$, $K > 1$, and $Q > 1$ if $\|u\|_p < K$, then there exist a residual (error) function y and a function z such that

$$\|z\|_p \leq MQ^{\nu} \tag{1.25}$$

$$\|y\|_0 \leq MQ^{-\mu} K^{\sigma} \tag{1.26}$$

and

$$F'(u)z + Fu - F'(u)u + fu + y = 0 \tag{1.27}$$

1.5 LEMMA The linearized equation (1.19) admits approximate solutions of order (μ,ν,σ) with $\sigma = \nu$.

Proof: The proof follows from (1.24) with η as in (1.23), where ν will be determined later (see Lemma 1.8). $\blacksquare$

Thus we have shown that if $\bar{z}$ is a solution of equation (1.9), then $z = S_\theta \bar{z}$ is an approximate solution of order (μ,ν,σ) with $\sigma = \nu$ of the linearized equation (1.19); that is, there exists a constant M and a function z which satisfy the relations (1.25) to (1.27).

Now put $z = u + h$. Then it follows from (1.27) that h is a solution of the equation

$$F'(u)h + Pu + y = 0 \tag{1.28}$$

and we get from (1.25), (1.26), and (1.12) with $g = Px + y$ that

$$\|h\|_0 \le C(\|Pu\|_0 + MQ^{-\mu}K^{\sigma}) \tag{1.29}$$

$$\|h\|_p \le \|u\|_p + MQK^{\nu} \tag{1.30}$$

where $\sigma = \nu$.

1.6 LEMMA Let $u \in V_0$ with $\|u\|_p < K$, $K > 1$, and with q_0 as in (1.8) let

$$q_0 < q < 1 \qquad \text{and} \qquad \bar{q} = \frac{q - q_0}{2} < q \tag{1.31}$$

If h is a solution of the equation (1.28), then there exists $0 < \varepsilon \le 1$ such that

$$\|P(u + \varepsilon h) - (1 - \varepsilon)Px\|_0 \le \varepsilon q\|Px\|_0 \tag{1.32}$$

and

$$\|h\|_0 \le (1 + \bar{q})\|Px\|_0 \tag{1.33}$$

Proof: For the proof see Lemma 1.1, Chapter 1. ∎

1.7 LEMMA For $u_n \in V_0$ let

$$\|u_n\|_p < A\exp(\alpha(1 - q)t_n) = K_n \tag{1.34}$$

where $t_n > 0$, and let α, A, $\bar{q}$, and q be such that

$$\max([\mu(1 - \nu) - \nu]^{-1}, (1 - q)^{-1}) < \alpha \qquad \text{and} \qquad \mu(1 - \nu) - \nu > 0 \tag{1.35}$$

with μ,ν as in Lemma 1.5 and

$$M^{1+1/\mu}(\bar{q}p_0)^{-1/\mu} < A[\alpha(1 - q) - 1] \qquad \alpha < \frac{p - s}{s} \tag{1.36}$$

where $p_0 = \|Pu\|_0$, and q,q_0 satisfy (1.31) and (1.8). Then there exists a number $Q = Q_n$ with $MQ_n^{-\mu}K^{\nu} < \bar{q}p_0 \exp(-(1 - q)t_n)$ and

$$\|u_n + \varepsilon h_n\|_p < A \exp(\alpha(1 - q)(t_n + \varepsilon)) \tag{1.37}$$

for all $0 < \varepsilon \leq 1$, where h_n is a solution of equation (1.28) with $u = u_n$, $y = y_n$, provided that (1.29) and (1.30) hold with $u = u_n$, $h = h_n$, $K = K_n$, and $Q = Q_n$.

Proof: For the proof see Lemma 1.2, Chapter 1. ∎

1.8 LEMMA There exist μ and ν such that

$$\bar{\nu} < \nu < 1 \qquad \mu(1 - \nu) - \nu > 0 \tag{1.38}$$

which satisfy the inequalities

$$\begin{aligned} &\theta^{p-p_\eta-1/2}K^{\bar{\nu}} < QK^{\nu} \\ &\theta^{-(m_2-m_1)}\eta^{1/2}K^{\bar{\nu}} < Q^{-\eta}K^{\nu} \end{aligned} \tag{1.39}$$

where $\bar{\nu} = (\bar{p} - \rho)/(p - \rho)$, $0 < \rho < m_2$ to be determined, and

$$m_2 - m_1 < m_3 < m_1 \tag{1.40}$$

Proof: For the proof see Lemma 1.6, Chapter 2. ∎

In the same way as in Chapter 2, we construct the following method of contractor directions (see also Altman, 1983a, b),

$$u_{n+1} = u_n + \varepsilon_n h_n \qquad t_{n+1} = t_n + \varepsilon_n \qquad t_0 = 0 \tag{1.41}$$

where h_n is a solution of equation (1.28) with $u = u_n$, $y = y_n$, and induction assumptions

$$\|u_n\|_p < A \exp(\alpha(1 - q)t_n) = K_n \tag{1.42}$$

$$\|Pu_n\|_0 \leq \|Pu_0\|_0 \exp(-(1 - q)t_n) \tag{1.43}$$

where α, A, and q are subject to (1.35) and (1.36) with ν satisfying the assumptions of Lemma 1.8.

The following estimates can be derived from (1.29) to (1.33),

$$\|h_n\|_0 \le 2C_0\|Pu_0\|_0 \exp(-(1 - q)t_n) \tag{1.44}$$

$$\|h_n\|_p \le \alpha(1 - q)A \exp(\alpha(1 - q)t_n) \tag{1.45}$$

Since

$$\|h_n\|_s \le C\|h_n\|_0^{1-\bar{s}}\|h_n\|_p^{\bar{s}} \qquad \bar{s} = \frac{s}{p}$$

we get from (1.44) and (1.45) (see Chapter 2),

$$\|h_n\|_s \le CN \exp(-\lambda(1 - q)t_n) \tag{1.46}$$

where $N = (2C\|Pu_0\|_0)^{1-\bar{s}}[\alpha(1 - q)A]^{\bar{s}}$ provided

$$\lambda = 1 - \bar{s}(1 + \alpha) > 0 \qquad \text{or} \qquad \alpha < \frac{1 - \bar{s}}{\bar{s}} \tag{1.47}$$

Hence, we obtain from (1.46) and (1.44)

$$\sum_{n=0}^{\infty} \varepsilon_n\|h_n\|_s \le CN[\lambda(1 - q)]^{-1} \exp(\lambda(1 - q)) = d_s$$

$$\sum_{n=0}^{\infty} \varepsilon_n\|h_n\|_0 \le 2C_0\|Pu_0\|_0(1 - q)^{-1} \exp(1 - q) = d_0$$

and consequently,

$$\sum_{n=0}^{\infty} \|u_{n+1} - u_n\|_{X_s} < r_0 \tag{1.48}$$

provided

$$d_0 + d_s < r_0 \tag{1.49}$$

1.1 THEOREM In addition to assumptions (A_1) to (A_3), suppose that $\|Pu_0\|_0$ is so small that condition (1.49) holds. Then equation (1.1) has a solution u such that $\|u_n - u\|_{X_s} \to 0$ as $n \to \infty$, where $u_n \subset V_0$ is determined by (1.41) with the same choice of $\{\varepsilon_n\}$ as in Chapter 2 and $\|u - u_0\|_{X_s} < r_0$.

Proof: The proof is the same as in Theorem 1.1, Chapter 2. ■

1.1 REMARK It follows from the proof of Lemma 1.2 that $\|v\|_{m_1}^2 < C\|g\|_{m_1}^2$, where $g = L_\eta v = \eta(-\Delta)^{m_3} v + F'(u)v$. Hence,

$$\|v\|_{m_1} < C\|L_\eta v\|_{m_1} < C\eta^{-\bar{k}}\|L_\eta v\|_{m_1} = C\eta^{-\bar{k}}\|g\|_{m_1}$$

where $0 < \eta < 1$ and $\bar{k} > 0$ are arbitrary. Thus $\bar{k}$ can be chosen so as to satisfy the condition of Lemma 1.8, i.e., $0 < \bar{k} < 1$.

2. NONLINEAR SYSTEMS OF FIRST-ORDER PARTIAL DIFFERENTIAL EQUATIONS

Moser (1966) investigated nonlinear systems of the form

$$F_k(x,u,u_x) = 0 \qquad \text{for } k = 1, 2, \ldots, m \tag{2.1}$$

where $F_k(x,u,p)$ are of period 2π in $x_1, \ldots, x_n$ and admit sufficiently many derivatives in

$$|u| + |p| \le 1$$

where p has $n \cdot m$ components p_{ki} corresponding to $\partial u_k/\partial x_i$, and the matrices $a^{(i)}$:

$$a_{kj}^{(i)}(x,u,p) = \frac{\partial F_k}{\partial p_{ji}} = \frac{\partial F_j}{\partial p_{ki}} \tag{2.2}$$

for $k,j = 1, \ldots, m$; $i = 1, \ldots, n$ are assumed to be symmetric, and denote by b:

$$b_{kj}(x,u,p) = \frac{\partial F_k}{\partial u_j}$$

To avoid boundary difficulties Moser investigated the differential equations (2.1) on the torus, and he assumed that all derivatives of $F = (F_k)$ up to a certain order $\ell = \ell(n)$ are bounded by a constant C for $|y| + |p| < 1$ and all x.

To prove an existence theorem for (2.1), Moser investigated the following linear positive symmetric system in the sense of Friedrichs,

$$Lu \equiv \sum_{i=1}^{n} a^{(i)} \frac{\partial}{\partial x_i} u + bu = g(x) \tag{2.3}$$

where $x = (x_1, \ldots, x_n)$, $u = (u_1, \ldots, u_m)$ is a vector, and $a^{(i)}(x)$, $b(x)$ are $m \times n$ matrices. Such a system is positive symmetric if all the matrices $a^{(i)}(x)$ are symmetric and

$$b + b^T - \sum_{i=1}^{n} a^{(i)}_{x_i} \tag{2.4}$$

is positive definite. Then the quadratic form

$$(u, Lu) = \int_\Omega \left\{ u \left(\sum_{i=1}^{n} a^{(i)} \frac{\partial}{\partial x_i} u \right) + ubu \right\} dx \tag{2.4'}$$

is positive definite if the integration is taken over

$$\Omega: \quad 0 \le x_i \le 2\pi$$

In order to construct smooth approximate solutions of the linearized equations for (2.1), Moser actually solved the following system

$$L_\eta w = (\eta^{2\alpha}(-\Delta)^\alpha + L)w = g \qquad 2\alpha = \rho = r - 1 \tag{2.5}$$

by using elliptic regularization with small $0 < \eta < 1$, and the following assumptions,

$$\|v\|_p^2 \le (Lv, v)_0 \tag{2.6}$$

$$\|v\|_p^2 \le c((Lv, v)_\rho + K_1^2 |v|_0^2) \qquad \rho = r - 1 \tag{2.7}$$

where $r - 1 > n/2$. He then proved that his degree of approximate linearization is $\mu = r - 1$ (in his case, $\nu = \sigma = 1$).

Moser also used his theorem on composition of functions $\phi \cdot v = \phi(x, v(x))$:

$$\|\phi \cdot v\|_r \le B(\|v\|_r + 1)$$

provided that all continuous derivatives of ϕ up to order r are bounded by B and $v \in X_r$ with $\max_x |v| = |v|_0 < 1$. Moreover, for $\phi(x, v(x), v_x)$:

$$\|\phi(x,v,v_x)\|_{r-1} \le B(\|v\|_r + 1)$$

holds, $x = (x_1,\ldots,x_n)$.

By using his well-known theorem, Moser proved the existence of a solution to (2.1).

In our approach based on Theorem 1.1, we make use of (1.10), and we assume the existence of a solution of equation (2.5) [see (1.9)] with $F'(u) = L$ and $-g = Fu - F'(u)u + fu$. By making use of Moser's theorem on composition of functions one can easily see that the assumptions (1.4) to (1.7) are verifiable.

Moser's estimate for $F(u,p) \equiv F(x,u,p)$

$$Q = F(u + v,\ P + v) - F(u,p) - F_u v = F_p v_x$$

$$|Q| \le c(|v| + |v_x|)^2 \tag{2.8}$$

where $|u|$, $|u_x| < \delta < 1$, $F = (F_1,\ldots,F_m)$, is rather essential for the applicability of his theorem to problem (2.1).

Now instead of (2.1) let us consider the following nonlinear system,

$$F_k(x,u,u_x) + f_k(x,u) = 0 \tag{2.9}$$

where $f(u) = (f_1(\cdot,u),\ldots,f_m(\cdot,u))$, $f_k(u) = f_k(\cdot,u)$ $(k = 1,\ldots,m)$. It is clear that Moser's theorem is not applicable to (2.9) even if f satisfies the same smoothness conditions as F, which is the same as in (2.1), and satisfies Moser's assumptions. For there still is a possibility that replacing the abstract derivative operator $L = F'$ by $F' + f'$ might result in violation of the assumption that (2.4) is positive definite. On the other hand, Theorem 1.1 shows the possibility of adding such an extra semilinear locally Lipschitzian term like f in (2.9). This is due to the fact that the linearized equation contains the same L [see (1.13)] but g changes.

Assumption (1.3) can be satisfied with appropriate choice of

$$1 + \frac{n}{2} \le s < m_2$$

by virtue of (2.8). Also $m_1 + m_3 < p$ should be satisfied. Finally, condition (1.49) is satisfied if $\|Pu_0\| = p_0$ is sufficiently small. One can put $u_0 = 0$. Then $p_0 = \|F(\cdot,9,0) + f(\cdot,0,0)\|_0$, and the smallness of p_0 can be replaced by that of $\max_x|F(x,0,0) + f(x,0)|$. Thus we have shown that the hypotheses of Theorem 1.1 are verifiable.

2.1 REMARK Remark 1.1 also applies to the problem (2.9).

5

Nonlinear Equations in a Scale of Hilbert-Sobolev Spaces, III

A further development of the general theory of global linearization iterative methods (GLIM) is presented in this chapter, and a method of elliptic regularization of degree $\bar{k} = 1/2$ is proposed which is based on GLIM-III. It turns out that under a somewhat weaker condition this method of elliptic regularization will be of degree $\bar{k} = 1$. Since then GLIM-I (or GLIM-II) is no longer applicable, one has to use GLIM-III, which has an advantage in this case. However, the class of nonlinear operator equations treated by this method is not as large as the one discussed in Chapter 4, since we assume here that $f \equiv 0$.

1. A NONLINEAR EXISTENCE THEOREM

Moser (1966) used the following Sobolev spaces X_p $(p = 0,1,\ldots)$ of real functions $u(x)$, $x = (x_1,\ldots,x_n)$ of period 2π in each $0 \le x_i \le 2\pi$ with inner product

$$(u,v)_p = \int u \cdot (-\Delta)^p v \, dx$$

where $\Delta = \Sigma_{i=1}^n (\partial/\partial x_i)^2 v \, dx$ is the Laplacean operator, integration is taken over $0 \le x_i \le 2\pi$, and dx is the volume element $dx_i \cdots dx_n$. Then $\|u\|_p = (u,u)_p^{1/2}$ is a seminorm which vanishes for constant functions if $p > 0$, but

$$\|u\|_{X_p} = (\|u\|_0^2 + \|u\|_p^2)^{1/2}$$

is a proper norm. The closure of all C^∞ functions (of period 2π) under this norm is a Hilbert-Sobolev space X_p. We know that $X_r \subset X_p$ if $0 \le p < r$ and $\|\cdot\|_n \le \|\cdot\|_m$ for $n \le m$. The following relation,

$$\|v\|_j \le \|v\|_k^{1-\lambda}\|v\|_i^\lambda$$

for $j = (1 - \lambda)k + \lambda i$; $0 \le \lambda \le 1$ will be needed and use will be made of a one-parameter family of smoothing operators S_θ, $\theta \ge 1$, in the sense of Nash (1956) (see also Moser, 1966). (For more details see Chapter 4.)

Let $W_0 \subset W_s$ (with s to be determined) be an open ball with center $u_0 \in X_p$ and radius $r_0 > 0$. Put $V_0 = W_0 \cap X_p$ and let V_s be the closure in X_s of V_0.

Consider the nonlinear operator equation

$$Pu = 0 \tag{1.1}$$

where P is closed on V_s and assumes its values in X_0. If the domain of P is $D \subset X_s$, then we put $V_0 = D \cap W_0 \cap X_p$.

We make the following assumptions.

(A_1) The operator P is differentiable, i.e., for each $u \in V_0$, there exists a linear operator $P'(u)$ such that for $h \in X_p$,

$$\varepsilon^{-1}\|P(u + \varepsilon h) - Pu - \varepsilon P'(u)h\|_0 \to 0 \quad \text{as } \varepsilon \to 0^+ \tag{1.2}$$

P' satisfies the following conditions:

$$\|P(u + h) - Pu - P'(u)h\|_0 \le C\|h\|_0^{2-\beta}\|h\|_p^\beta \tag{1.3}$$

for some $C > 0$, $0 \le \beta < 1$, and all $u \in V_0$

$$\|P'(u)h\|_0 \le C\|h\|_{m_1} \tag{1.4}$$

$$\|P'(u)h\|_{m_1} \le C\|h\|_{m_2} \tag{1.5}$$

$$\|Pu - P'(u)u\|_{m_1} \le C\|u\|_{m_2} \tag{1.6}$$

where $m_2 > m_1 > 0$.

(A_2) The modified linearized equation

$$\eta(-\Delta)^{m_3} v + P'(u)v + Pu - P'(u)u = 0 \qquad m_2 - m_1 \le m_3 \le m_1 \quad (1.7)$$

has a solution if $u \in V_0$ and $0 < \eta < 1$, and

$$0 \le (P'(u)v,v)_{m_1} \quad (1.8a)$$

or

$$\|v\|_0^2 \le (P'(u)v,v)_0$$
$$0 \le c[(P'(u)v,v)_{m_1} + K_1^2\|v\|_0^2] \quad (1.8b)$$

for some c, K_1, and all $u \in V_0$.

(A_3) If h is a solution of the equation

$$P'(u)h + g = 0 \quad (1.9)$$

where $u \in V_0$, then in case (1.8a) we have

$$\|h\|_0 \le C_0\|g\|_0 \quad (1.10)$$

for some $C_0 > 0$.

1.1 LEMMA If $\bar{z}$ is a solution of the equation

$$L_\eta = \eta(-\Delta)^{m_3}\bar{z} + P'(u)\bar{z} = g \quad (1.11)$$

then

$$\|\bar{z}\|_{\bar{p}} \le C\eta^{-\bar{k}}\|g\|_{m_1} \qquad \text{with} \qquad \bar{k} = 1,\ \bar{p} = m_3 + m_1 \quad (1.12)$$

Thus, $\bar{k}$, the degree of elliptic regularization, is equal to one.

Proof: We have by (1.8a)

$$\eta\|\bar{z}\|_{\bar{p}}^2 \le C\|g\|_{m_1}\|\bar{z}\|_{m_1},$$

Hence, relation (1.12) follows. In case (1.8b), we have

$$\eta\|\bar{z}\|_{\bar{p}}^2 \le C[(L_\eta\bar{z},\bar{z})_{m_1} + K_1^2\|\bar{z}\|_0^2]$$

$$\le C\|g\|_{m_1}\|\bar{z}\|_{m_1} + CK_1^2\|g\|_0\|\bar{z}\|_0$$

since

$$\eta\|\bar{z}\|_{\bar{p}}^2 + \|\bar{z}\|_0^2 \le (L_\eta\bar{z},\bar{z})_0 = (g,\bar{z})_0 \le 2^{-1}(\|g\|_0^2 + \|\bar{z}\|_0^2)$$

and relation (1.12) follows. ∎

For $u \in V_0$, let $\bar{z}$ be a solution of equation (1.7). Then we have:

1.2 LEMMA The following estimates hold for $u \in V_0$ with $\|u\|_p < K$, $K > 1$.

$$\|P'(u)(I - S_\theta)\bar{z}\|_0 \le M_1\theta^{-(m_2-m_1)}\eta^{-1}K^{\bar{\nu}} \tag{1.13}$$

$$\|z\|_p \equiv \|S_\theta\bar{z}\|_p \le M_2\theta^{p-\bar{p}}\eta^{-1}K^{\bar{\nu}} \tag{1.14}$$

for some $M_1, M_2 > 0$ and $z = S_\theta\bar{z}$, $\bar{\nu} = (m_2 - \rho)/(p - \rho)$, with ρ to be determined.

Proof: We get from (1.4),

$$\|P'(u)(I - S_\theta)\bar{z}\|_0 \le C\|(I - S_\theta)\bar{z}\|_{m_1} \le C\theta^{-(m_2-m_1)}\|\bar{z}\|_{m_2}$$

Since

$$\|\bar{z}\|_{m_2} \le \|\bar{z}\|_{\bar{p}} \le C\eta^{-1}\|Pu - P'(u)u\|_{m_1} \le C\eta^{-1}\|u\|_{m_2}$$

by (1.6) and

$$\|u\|_{m_2} \le C\|u\|_s^{1-\bar{\nu}}\|u\|_p^{\nu} \le Cr_0^{1-\bar{\nu}}K^{\bar{\nu}} \tag{1.15}$$

where $\bar{\nu} = (m_2 - \rho)/(p - \rho)$ with $0 < \rho < m_2$ to be determined, we obtain

$$\|\bar{z}\|_{\bar{p}} \le C\eta^{-1}K^{\bar{\nu}} \tag{1.16}$$

and (1.13) follows.

We have

$$\|z\|_p \le C\theta^{p-\bar{p}}\|\bar{z}\|_{\bar{p}}$$

and (1.14) results from (1.16). ∎

Consider the linearized equation

$$P'(u)z + Pu - P'(u)u = 0 \tag{1.17}$$

and put $z = S_\theta \bar{z}$, where $\bar{z}$ is a solution of equation (1.7) satisfying

$$\|\bar{z}\|_{\bar{p}} \le C\eta^{-1}K^{\bar{\nu}} \qquad \bar{p} = m_3 + m_1 \tag{1.18}$$

for some $C > 0$, by virtue of (1.12), (1.6), and (1.15).

1.3 LEMMA The following estimate holds,

$$\|P'(u)z + Pu - P'(u)u\|_0 \le M_2(\theta^{-m} + \eta)\eta^{-1}K^{\bar{\nu}} \tag{1.19}$$

for $u \in V_0$ with $\|u\|_p < K$ and $m = m_2 - m_1$.

Proof: We have by (1.13),

$$\begin{aligned}\|P'(u)z + Pu - P'(u)\|_0 &\le \|\eta(-\Delta)^{m_3}\bar{z} + P'(u)\bar{z} + Pu - P'(u)u\|_0 \\ &\quad + \|P'(u)(I - S_\theta)\bar{z}\|_0 + \eta\|(-\Delta)^{m_3}\bar{z}\|_0 \\ &\le M_1\theta^{-m}\eta^{-1}K^{\bar{\nu}} + \eta\|\bar{z}\|_{2m_3}\end{aligned}$$

where $2m_3 \le m_3 + m_1 = \bar{p}$ and (1.19) results by (1.16).

Now let $0 < \eta < 1$ and put

$$Q = \theta^{p-\bar{p}} \qquad \text{and} \qquad \mu = \frac{m_2 - m_1}{p - \bar{p}} \tag{1.20}$$

and let η be such that

$$\eta^{-1}K^{\bar{\nu}} < K^{1+\alpha} \qquad \eta < Q^{-\mu},\ \nu = 1 + \alpha \tag{1.21}$$

for some α to be determined with $0 < \alpha < 1$. Then relation (1.19) can be written as

$$\|P'(u)z + Pu - P'(u)u\|_0 \le 2M_2Q^{-\mu}K^{\nu} \tag{1.22}$$

since $\theta^{-(m_2-m_1)} = Q^{-\mu}$ by (1.20). ■

1.1 DEFINITION Let $\mu > 0$, $\nu \ge 0$, $\sigma \ge 0$ be given numbers. Then the linearized equation (1.17) admits approximate solutions of order (μ,ν,σ) if there exists a constant $M > 0$ with the following property. For every $u \in V_0$, $K > 1$, and $Q > 1$ if $\|u\|_p < K$, then there exist a residual (error) function y and a function z such that

$$\|z\|_p \le MQK^{\nu} \tag{1.23}$$

$$\|y\|_0 \le MQ^{-\mu}K^{\sigma} \tag{1.24}$$

and

$$P'(u)z + Pu - P'(u)u + y = 0 \tag{1.25}$$

1.4 LEMMA The linearized equations (1.17) admit approximate solutions of order (μ,ν,σ) with $\sigma = \nu$.

Proof: The proof follows from (1.22) with ν as in (1.21), (1.23), and (1.24). ■

Thus we have shown that if $\bar{z}$ is a solution of equation (1.7), then $z = S_\theta\bar{z}$ is an approximate solution of order (μ,ν,σ) with $\nu = \sigma$ of the linearized equation (1.17); that is, there exists a constant $M > 0$ and a function z which satisfy the relations (1.23) to (1.25).

Now put $z = u + h$. Then it follows from (1.25) that h is a solution of the equation

$$P'(u)h + Pu + y = 0 \tag{1.26}$$

and we get from (1.23), (1.24), and (1.10) with $g = Pu + g$ that

$$\|h\|_0 \le C(\|Pu\|_0 + MQ^{-\mu}K^{\sigma}) \tag{1.27}$$

$$\|h\|_p \leq \|u\|_p + MQK^{\sigma} \tag{1.28}$$

where $\sigma = \nu = 1 + \alpha$.

(A_4) Let α, τ; $\lambda < \mu$, be such that

(i) $1 < (\mu - \lambda)^{-1}(1 + \alpha(1 + \lambda) + \mu) < \tau < 2 - \alpha_0 < 2 - \alpha$

(ii) $0 < 2\lambda < [\mu(1 - \alpha_0) - (1 + \alpha_0)]/(1 + \alpha_0)$

where $0 < \alpha < \alpha_0 < 1$ is such that

(iii) $\mu > (1 + \alpha_0)/(1 - \alpha_0) > 1$

and let β be such that

(iv) $0 < \beta < \mu\lambda(\alpha_0 - \alpha)[(1 + \alpha)(1 + \mu) + \lambda(2 + \mu)]^{-1} < 1$

As in Chapter 1, one can put $\alpha_0 = (\mu - 1)/(\mu + 3)$ in (i)-(iv). Then instead of (ii), we have

(iia) $0 < 2\lambda < (\mu - 1)/(\mu + 1)$

In the same way as in Chapter 1, we construct the following iterative method,

$$u_{n+1} = z_n = u_n + h_n \qquad n = 0, 1, \dots \quad (u_n \in V_0)$$

where z_n is a solution of the equation (1.25) with $u = u_n$, $y = y_n$, and Q_n in (1.23) and (1.24) is chosen so as to satisfy the induction assumptions

$$\|u_n\|_p < K_n \tag{1.30}$$

$$\|Pu_n\|_0 < K_n^{-\lambda} \tag{1.31}$$

where $K_{n+1} = K_n^{\tau+\alpha}$, and λ, α, and τ satisfy assumption (A_4).

1.1 THEOREM Suppose that assumptions (A_1) to (A_4) are satisfied. Then there exists a constant $K_0(M,\beta,\mu,\lambda,\alpha) > 1$ such that if

$$\|Pu_0\|_0 < K_0 \qquad \|u_0\|_p < K_0 \qquad C(2C_0)^{1-s/p}\,\Sigma_{n=0}^{\infty} K_n^{-\delta} < r_0$$

hold, where $\delta = (1 - s/p)\lambda - \tau s/p > 0$, provided

$$\frac{s}{p} < \frac{\lambda}{\lambda + 2}$$

and $K_{n+1} = K_n^{\tau+\alpha}$ with τ,α as in (i), then equation (1.1) has a solution u which is a limit of $\{u_n\}$ and

$$\|u_n - u\|_s \to 0 \qquad \text{as } n \to \infty$$

where $\{u_n\}$ is determined by (1.29).

Proof: The proof is the same as in Chapter 1. ∎

2. NONLINEAR SYSTEMS OF PARTIAL DIFFERENTIAL EQUATIONS

Moser (1966) investigated nonlinear systems of the form

$$F_k(x,u,u_x) = 0 \qquad \text{for} \qquad k = 1, 2, \ldots, m \tag{2.1}$$

where $F_k(x,u,p)$ are of period 2π in $x_1, \ldots, x_n$ and admit sufficiently many derivatives in

$$|u| + |p| \le 1$$

where p has $n \cdot m$ components p_{ki} corresponding to $\partial u_k/\partial x_i$, and the matrices $a^{(i)}$:

$$a_{k_j}^{(i)}(x,u,p) = \frac{\partial F_k}{\partial p_{ji}} = \frac{\partial F_j}{\partial p_{ki}} \tag{2.2}$$

for $k,j = 1, \ldots, m$ and $i = 1, \ldots, n$ are assumed to be symmetric. Denote by b:

$$b_{kj}(x,u,p) = \frac{\partial F_k}{\partial u_j}$$

To avoid boundary difficulties Moser investigated the differential equations (2.1) on the torus, and he assumed that all derivatives of $F = (F_k)$ up to a certain order $\ell = \ell(n)$ are bounded by a constant C for $|y| + |p| < 1$ and all x.

To prove an existence theorem for (2.1), Moser investigated the following linear positive symmetric system in the sense of Friedrichs (1954),

$$Lu \equiv \sum_{i=1}^{n} a^{(i)} \frac{\partial}{\partial x_i} u + bu = g(x) \tag{2.3}$$

where $x = (x_1,\ldots,x_n)$, $u = (u_1,\ldots,u_m)$ is a vector, and $a^{(i)}(x)$, $b(x)$ are $m \times n$ matrices. Such a system is positive symmetric if all the matrices $a^{(i)}(x)$ are symmetric and

$$b + b^T - \sum_{i=1}^{n} a^{(i)}_{x_i} \tag{2.4}$$

is positive definite. Then the quadratic form

$$(u,Lu) = \int_{\Omega} \left\{ u\left(\sum_{i=1}^{n} a^{(i)} \frac{\partial}{\partial x_i} u \right) + ubu \right\} dx \tag{2.4'}$$

is positive definite if the integration is taken over

$$\Omega: \quad 0 \le x_i \le 2\pi$$

In order to construct smooth approximate solutions of the linearized equations for (2.1), Moser actually solved the following system,

$$L_\eta w = (\eta^{2\alpha}(-\Delta)^{\alpha} + L)w = g \qquad 2\alpha = p = r - 1 \tag{2.5}$$

by using elliptic regularization with small $0 < \eta < 1$, and the following assumptions:

$$\|v\|_0^2 \le (Lv,v)_0 \tag{2.6}$$

$$\|v\|_p^2 \le C((Lv,v)_p + K_1^2|v|_0^2) \qquad p = r - 1 \tag{2.7}$$

where $r - 1 > n/2$. He then proved that his degree of approximate linearization is $\mu = r - 1$ (in his case, $\nu = \sigma = 1$).

Moser also used his theorem on composition of functions $\varphi \circ v = \varphi(x,v(x))$:

$$\|\varphi \circ v\|_r \le B(\|v\|_r + 1)$$

provided that all continuous derivatives of φ up to order r are bounded by B and $v \in X_r$ with $\max_x |v| = |v|_0 < 1$. Moreover, for $\varphi(x, v(x) v_x)$,

$$\|\varphi(x, v, v_x)\|_{r-1} \le B(\|v\|_r + 1)$$

holds, $x = (x_1, \ldots, x_n)$.

By using his well-known theorem Moser proved the existence of a solution to (2.11).

In our approach based on Theorem 1.1 we make use of (1.10), and we assume the existence of a solution of the equation (2.5) [see (1.9)] with $F'(u) = L$ and $-g = Fu - F'(u)u$. Thus our assumptions (A_2) are to replace Moser's conditions (2.6) and (2.7).

By making use of Moser's theorem on composition of functions, one can easily see that the assumptions (1.4) to (1.7) are verifiable.

Moser's estimate for $F(u,p) \equiv F(x,u,p)$,

$$|Q| \le C(|v| + |v_x|)^2$$

where $Q = F(u + v, p + v) - F(u,p) - F_u v - F_p v_x$, yields (1.3) and is essential in our case as well as in Moser (1966).

Thus we have seen that the hypotheses of Theorem 1.1 are verifiable.

6

Elliptic Regularization Without Smoothing Operators

Approximate solutions of the linearized equations employed in Chapter 1 can be constructed without making use of smoothing operators. But then stronger conditions have to be imposed on the regularizing linear operator L which is involved in the modified linearized equations. In other words, a "gain of derivatives" is required for the estimate of the solutions of that linearized equation, i.e., a stronger norm involving more derivatives. Then the general convergence and existence scheme proposed in Chapter 1 is applied to each global linearization iterative method (GLIM).

1. ELLIPTIC REGULARIZATION AND GLIM-I

Let $Z \subset E \subset Y \subset X$ be Banach spaces with norms $\|\cdot\|_X \leq \|\cdot\|_Y \leq \|\cdot\|_E \leq \|\cdot\|_Z$ and let $W_0 \subset Y$ be an open ball with center $x_0 \in Z$ and radius $r_0 > 0$. Put $V_0 = W_0 \cap Z$ and let V_1 be $\|\cdot\|_Y$-closure of V_0. As in Chapter 1, consider the operator equation

$$Px \equiv Lx + Fx + fx = 0 \tag{1.1}$$

where $P\colon\ V_1 \to X$. If the domain of P is $\mathcal{D}$, then we put $V_0 = \mathcal{D} \cap W_0 \cap Z$.

We make the following assumptions.

(A_0) There exist constants $C > 0$, $0 < \bar{s} < 1$, such that

$$\|x\|_Y \leq C\|x\|_X^{1-\bar{s}}\|x\|_Z^{\bar{s}}$$

(A_1) As in Chapter 1, we assume that F is differentiable and F' satisfies condition (1.6), Chapter 1.

P is closed on V_1.

(A_2) f is locally Lipschitzian and satisfies condition (1.9), Chapter 1.

(A_3) There exists a linear (regularizing) operator $L = L(\eta)$ such that

$$\|Lx\|_X \leq C_1\|x\|_Z \tag{1.2}$$

for some constant $C_1 > 0$. The linearized equation

$$Lz + F'(x)z + \eta Lz + Fx - F'(x)x + fx = 0 \qquad 0 < \eta < 1 \tag{1.3}$$

has a solution z such that

$$\|z\|_Z \leq C\eta^{-1/\bar{\mu}}\|Fx - F'(x)x + fx\|_E \tag{1.4}$$

for some $C > 0$, $\bar{\mu} > 1$ and all $x \in V_0$.

(A_4) There exist constants $C > 0$, $0 < \nu < 1$ such that $x \in V_0$ implies

$$\|Fx - F'(x)x + fx\|_E \leq C\|x\|_Z^{\nu} \tag{1.5}$$

(A_5) For $x \in V_0$ if h is a solution of the equation

$$Lh + F'(x)h + g = 0 \tag{1.6}$$

then

$$\|h\|_X \leq C_0\|g\|_X$$

for some constant C_0 such that $C_0q_0 < 1$ with q_0 as in (1.9), Chapter 1.

1.1 REMARK Suppose that instead of (1.2) we have

$$\|Lx\|_X \leq C\|x\|_E$$

and there exist $C > 0$, $0 < \sigma < 1$ such that

$$\|x\|_E \le C\|x\|_X^{1-\sigma}\|x\|_Z^{\sigma}$$

then relation (1.2) holds for $x \in V_0$. Also suppose that instead of (1.5) we have

$$\|Fx - F'(x)x + fx\|_E \le C\|x\|_V$$

for some $C > 0$ and $E \subset V \subset Z$, and there exist $C > 0$, $0 < \nu < 1$ such that

$$\|x\|_V \le C\|x\|_X^{1-\nu}\|x\|_Z^{\nu}$$

then relation (1.5) holds for $x \in V_0$.

1.1 LEMMA The linearized equation

$$Lz + F'(x)z + Fx - F'(x)x + fx = 0 \tag{1.8}$$

admits approximate solutions of order (μ,ν,σ) in the sense of Definition 1.1, Chapter 1.

Proof: Let z be a solution of the equation (1.3) with $x \in V_0$, $\|x\|_Z < K$, $K > 1$, where $\eta = Q^{-\bar{\mu}}$ for $Q > 1$. Then relations (1.4) and (1.5) imply

$$\|z\|_Z < MQK^{\nu} \tag{1.9}$$

where M is some constant. On the other hand, we get from (1.3),

$$\|Lz + F'(x)z + Fx - F'(x)x + fx\|_X \le \eta\|Lx\|_X$$
$$\le C_1Q^{-\mu}\|x\|_Z^{\sigma} \le MQ^{-\mu}K^{\sigma} \qquad \mu = \bar{\mu} - 1$$

Thus $K > 1$, $Q > 1$, and $x \in V_0$ with $\|x\|_Z < K$ imply the existence of z,y such that

$$\|z\|_Z < MQK^{\nu} \tag{1.10}$$

and

$$\|y\|_X \le MQ^{-\mu}K^{\sigma} \tag{1.11}$$

for some constant $M > 0$, and

$$Lz + F'(x)z + Fx - F'(x)x + fx + y = 0 \tag{1.12}$$

where z is a solution of the equation (1.3) with $\eta = Q^{-\bar{\mu}}$. ■

Now put z = x + h. Then we get from (1.12),

$$Lh + F'(x)h + Px + y = 0 \tag{1.13}$$

Next we can construct the following iterative method of contractor directions in the same way as in Chapter 1. Put

$$x_{n+1} = x_n + \varepsilon_n h_n \qquad t_{n+1} = t_n + \varepsilon_n \qquad t_0 = 0 \tag{1.14}$$

where h_n is a solution of the equation (1.13) with $x = x_n$, $y = y_n$, and $x_n \in V_0$. The induction assumptions for $\{x_n\}$ are the same as in Chapter 1, i.e., (1.27) and (1.28), where $0 < \bar{q} < q < 1$ satisfy relations (1.13), and $\alpha > 1$, $A > 0$ satisfy (1.18) to (1.20), and (1.33), Chapter 1. The choice of $\{\varepsilon_n\}$ is exactly the same as in Chapter 1 and its estimate (1.34) is valid.

1.1 THEOREM In addition to the assumptions (A_0) to (A_5), suppose that conditions (1.13) and (1.18) to (1.20), Chapter 1, hold, and $p_0 = \|Px_0\|_X$ is so small that relation (1.36) is satisfied. Then the equation (1.1) has a solution x which satisfies (1.37), Chapter 1, where $\{x_n\}$ is determined by (1.14).

Proof: The proof of Theorem 1.1, Chapter 1, carries over. ■

2. ELLIPTIC REGULARIZATION AND GLIM-II

Approximate solutions of the linearized equations discussed in Chapter 1 can also be constructed without making use of smoothing operators. Let $Z \subset E \subset Y \subset X$ be Banach spaces with norms $\|\cdot\|_X \le \|\cdot\|_Y \le \|\cdot\|_E \le \|\cdot\|_Z$ and let $W_0 \subset Y$ be an open ball with center $x_0 \in Z$ and radius $r_0 > 0$. Put $V_0 = W_0 \cap Z$ and let V_1 be the $\|\cdot\|_Y$-closure of V_0. As in Section 1, consider the operator equation

$$Px \equiv Ax + Fx + fx = 0 \tag{2.1}$$

where $P\colon\ V_1 \to X$. If the domain of P is D, then we put $V_0 = D \cap W_0 \cap Z$.

We make the following assumptions.

(A_0) $\|x\|_Y \le C\|x\|_X^{1-\bar{s}}\|x\|_Z^{\bar{s}}$

for some constants $C > 0$, $0 < \bar{s} < 1$.

(A_1) As in Chapter 1, we assume that F is differentiable and satisfies (2.6), Chapter 1.

P is closed on V_1.

(A_2) f satisfies condition (2.9), Chapter 1.

(A_3) There exists a linear (regularizing) operator $L = L(\eta)$ such that

$$\|Lx\|_X \le C\|x\|_Z \tag{2.2}$$

and the linearized equation

$$Az + F'(x)z + \eta Lz + Fx - F'(x)x + fx = 0 \qquad 0 < \eta < 1 \tag{2.3}$$

has a solution z such that

$$\|z\|_Z \le C\eta^{-1/\bar{\mu}}\|Fx - F'(x)x + fx\|_E \tag{2.4}$$

for some $C > 0$, $\mu > 1$, and all $x \in V_0$.

(A_4) There exists constants $C > 0$, $0 < \nu < 1$ such that $x \in V_0$ implies

$$\|Fx - F'(x)x + fx\|_E \le C\|x\|_Z^{\nu} \tag{2.5}$$

(A_5) For $x \in V_0$ if h is a solution of the equation

$$Ah + F'(x)h + g = 0 \tag{2.6}$$

then

$$\|h\|_X \le C_0\|g\|_X \tag{2.7}$$

for some constant C_0 subject to (2.13), Chapter 1.

2.1 REMARK Suppose that instead of (2.2) we have

$$\|Lx\|_X \le C\|x\|_E$$

and there exist $C > 0$, $0 < \sigma < 1$ such that

$$\|x\|_E \le C\|x\|_X^{1-\sigma}\|x\|_Z^{\sigma}$$

then relation (2.2) holds for $x \in V_0$. Also suppose that instead of (2.5) we have

$$\|Fx - F'(x)x + fx\|_E \le C\|x\|_V$$

for some $C > 0$ and $E \subset V \subset Z$, and there exist $C > 0$, $0 < \nu < 1$ such that

$$\|x\|_V \le C\|x\|_X^{1-\nu}\|x\|_Z^{\nu}$$

Then relation (2.5) holds for $x \in V_0$.

2.1 LEMMA The linearized equation

$$Az + F'(x)z + Fx - F'(x)x + fx = 0 \tag{2.8}$$

admits approximate solutions of order (μ,ν,σ) in the sense of Definition 1.1, Chapter 1.

Proof: Let z be a solution of the equation (2.3) with $x \in V_0$, $\|x\|_Z < K$, $K > 1$, where $\eta = Q^{-\bar{\mu}}$ for $Q > 1$. Then relations (2.4) and (2.5) imply

$$\|z\|_Z \le MQK^{\nu} \tag{2.9}$$

where $M > 0$ is some constant. On the other hand, we get from (2.3)

$$\|Az + F'(x)z + Fx - F'(x)x + fx\|_X = \eta\|Lz\|_X$$

$$\le C_1Q^{-\mu}\|x\|_Z^{\sigma} \le MQ^{-\mu}K^{\sigma}$$

where $\mu = \bar{\mu} - 1$. Thus $K > 1$, $Q > 1$, and $x \in V_0$ with $\|x\|_Z < K$ imply the existence of z,y satisfying (2.9) and

$$\|y\|_X \le MQ^{-\mu}K^{\sigma} \tag{2.10}$$

for some constant $M > 0$, and

$$Az + F'(x)z + Fx - F'(x)x + fx + y = 0 \tag{2.11}$$

where z is a solution of the equation (2.3) with $\eta = Q^{-\mu}$.

Now put $z = x + h$. Then we get from (2.11)

$$Ah + F'(x)h + Px + y = 0 \tag{2.12}$$

Next we can construct the following iterative method in the same way as in Chapter 1. Put

$$x_{n+1} = z_n = x_n + h_n \tag{2.13}$$

where z_n, h_n are solutions of the equation (2.11) with $x = x_n$, $y = y_n$, and (2.12), respectively, the induction assumptions for $\{x_n\}$ are the same as in Chapter 1, i.e., (2.18) and (2.19), where $\bar{q}$ and q satisfy condition (2.17), and α, A satisfy (2.33) and (2.21); the estimate (2.29) is also valid. ■

2.1 THEOREM In addition to the assumptions (A_0) to (A_5), suppose that conditions (2.17), (2.33), (2.21), and (2.35), Chapter 1, with N as in (2.29) are satisfied. Then the equation (2.1) has a solution x which satisfies (2.36), Chapter 1, where $\{x_n\}$ is determined by (2.13).

Proof: The proof of Theorem 1.1, Chapter 1, carries over. ■

3. ELLIPTIC REGULARIZATION AND GLIM-III

Approximate solutions of the linearized equations discussed in Chapter 1 can also be constructed without making use of smoothing operators. Let $Z \subset Y \subset X$ and $E \subset G$ be Banach spaces with norms $\|\cdot\|_Z \geq \|\cdot\|_Y \geq \|\cdot\|_X$ and $\|\cdot\|_E \geq \|\cdot\|_G$. Let $W_0 \in Y$ be an open ball with center $x_0 \in Z$ and radius $r_0 > 0$. Put $V_0 = W_0 \cap Z$ and let V_1 be the $\|\cdot\|_Y$-closure of V_0. As in Section 2, consider the operator equation

$$Px \equiv Ax + Fx = 0 \tag{3.1}$$

where $P\colon V_1 \to G$. If the domain of P is D, then put $V_0 = D \cap W_0 \cap Z$.

We make the following assumptions:

(A_0) $\|x\|_Y \leq C\|x\|_X^{1-\bar{s}}\|x\|_Z^{\bar{s}}$ for some constants $C > 0$, $0 < \bar{s} < 1$.

(A_1) As in Section 1 we assume that F is differentiable and F' satisfies (3.6), Chapter 1.

P is closed on V_1.

(A_2) There exists a linear (regularizing) operator $L = L(\eta)$ such that

$$\|Lx\|_G \le C\|x\|_Z \tag{3.2}$$

and the linearized equation

$$Az + F'(x)z + \eta Lz + Fx - F'(x)x = 0 \tag{3.3}$$

$0 < \eta < 1$, has a solution z such that

$$\|z\|_Z \le C\eta^{-1/\bar{\mu}}K^{\alpha}\|Fx - F'(x)x\|_G \tag{3.4}$$

for some $C > 0$, $\bar{\mu} > 1$, and all $x \in V_0$ with $\|x\|_Z < K$; $0 < \alpha < 1$.

(A_3) There exists a constant $C > 0$ such that

$$\|Fx - F'(x)x\|_G \le C\|x\|_Z \tag{3.5}$$

for all $x \in V_0$.

(A_4) For $x \in V_0$, if h is a solution of the equation

$$Ah + F'(x)h + g = 0 \tag{3.6}$$

then

$$\|h\|_X \le C_0\|g\|_G \tag{3.7}$$

for some constant C_0.

3.1 LEMMA The linearized equation

$$Az + F'(x)z + Fx - F'(x)x = 0 \tag{3.8}$$

admits approximate solutions of order (μ,ν,σ) in the sense of Definition 1.1, Chapter 1, with $\nu = \sigma = 1 + \alpha$.

Proof: Let z be a solution of the equation (3.3) with $x \in V_0$, $\|x\|_Z < K$, $K > 1$, where $\eta = Q^{-\bar{\mu}}$ for $Q > 1$. Then relations (3.4) and (3.5) imply

$$\|z\|_Z \le MQK \tag{3.9}$$

where $M > 0$ is some constant. On the other hand, we get from (3.3)

$$\|Az + F'(x)z + Fx - F'(x)x\|_G = \eta\|Lx\|_G$$

$$\le C_1 Q^{-\mu}\|x\|_Z \le MQ^{-\mu}K \qquad \mu = \bar{\mu} - 1$$

Thus $K > 1$, $Q > 1$, and $x \in V_0$ with $\|x\|_Z < K$ imply the existence z,y satisfying (3.9) and

$$\|y\|_X \le MQ^{-\mu}K \tag{3.10}$$

for some constant $M > 0$, and

$$Az + F'(x)z + Fx - F'(x)x + y = 0 \tag{3.11}$$

where z is a solution of the equation (3.3) with $\eta = Q^{-\bar{\mu}}$. ∎

Now put $z = x + h$. Then we get from (3.11)

$$Ah + F'(x)h + Px + y = 0 \tag{3.12}$$

Next we can construct the following iterative method in the same way as in Chapter 1. Put

$$x_{n+1} = z_n = x_n + h_n \tag{3.13}$$

where z_n, h_n are solutions of the equation (3.11) with $x = x_n$, $y = y_n$, and (3.12), respectively. The induction assumptions for $\{x_n\}$ are the same as in Chapter 1; i.e., (3.15), (3.16); $K_{n+1} = K^{\tau+\alpha}$ with τ,α as in Chapter 1.

3.1 THEOREM In addition to (A_0)-(A_4), suppose that assumption (A_4) of Section 1 is satisfied. Then the statement of Theorem 3.1, Chapter 1, is true.

Proof: The proof is the same as that of Theorem 3.1, Chapter 1. ∎

A similar theorem can be proved by making use of Theorem 3.2, Chapter 1. Then we put $\alpha = 0$ everywhere and replace (A_4) of Section 3 by (A'_4).

The above theorems can be applied to nonlinear partial differential equations in the same way as in Moser (1966).

3.1 REMARK Definition 1.2 of approximate linearization (see Remark 1.1, Chapter 1) can also be applied to this chapter as follows.

Suppose that the linearized equation

$$Lz + F'(x)z + \eta Lz + g(x) = 0$$

has a solution z such that

$$\|z\|_Z \le C\eta^{-\bar{k}}\|g(x)\|_E$$

where $0 < \bar{k} < 1$, and

$$\|g(x)\|_E \le C\|x\|_Z^{\bar{\nu}} \qquad \text{and} \qquad \|Lx\|_X \le C\|x\|_Z$$

where $0 < \bar{\nu} < 1$. Hence we get

$$\|z\|_Z \le M\eta^{-\bar{k}}K^{\bar{\nu}} \qquad \text{and} \qquad \|y\|_X \le M\eta\eta^{-\bar{k}}K^{\bar{\nu}}$$

Then one can find $\bar{\nu} < \nu \le 1$ and $\sigma > 0$ such that

$$\eta^{-\bar{k}}K^{\bar{\nu}} < K^{\nu} \qquad \text{and} \qquad \eta\eta^{-\bar{k}}K^{\bar{\nu}} < K^{-\sigma}$$

In fact, it results from these inequalities that

$$K^{-(\nu-\bar{\nu})/\bar{k}} < \eta < K^{-(\sigma+\bar{\nu})/(1-\bar{k})}$$

Hence, ν,σ exist if $0 < \bar{k} < 1 - \bar{\nu}$ and

$$0 < (\sigma + \bar{\nu})/(1 - \bar{k}) < (\nu - \bar{\nu})/\bar{k}$$

or

$$0 < \sigma < (\nu - \bar{\nu})(1 - \bar{k})/\bar{k}\nu \qquad \text{and} \qquad \bar{\nu} < \nu \le 1$$

3.2 REMARK Moser's (1966) theorem requires, among other assumptions, that $\|x\|_Z < K$ $(K > 1)$ implies $\|Px\|_E < MK$ for some $M > 0$. But this cannot be satisfied, since A in (2.1) is nonbounded. If $A = 0$, then the estimate in (A_1) does not hold for $F + f$, f being Lipschitz. For this reason Moser's theorem is not applicable to the equation (2.1) or to the nonlinear evolution equation (1.1), Chapter 7.

7

Convex Approximate Linearization and Global Linearization Iterative Methods for Nonlinear Evolution Equations

Kato's famous theorem (1975) on existence of solutions of quasi-linear evolution equations was the first very important step in the theory of nonlinear evolutions equations which he successfully applied to partial differential equations of mathematical physics. Although his theorem is applicable only to reflexive Banach spaces, a generalization to nonreflexive Banach spaces due to a different method was given in Altman (1981). But the nonlinear problem remained open until three different mehtods were proposed in a series of investigations (Altman, 1984a,c,d) of the nonlinear case. All of them are based on a special kind of smooth approximate solutions of the linearized equations, a notion introduced by Nash (1956) to handle the so-called loss of derivatives, and developed by Moser (1966), Hörmander (1976), and others. The three methods mentioned above are: (1) an iterative method with small steps, which is a method of contractor directions; (2) an iterative method with step size equal to one; and (3) a rapidly convergent iterative method. The first two methods are related to earlier investigations (Altman, 1981; 1983a,b). The third one makes use of Moser's essential technique. Although Moser's method is designed for operator equations and is not applicable to nonlinear evolution equations, it is rather surprising that his essential iteration technique could be adapted under an entirely different set of hypotheses.

As mentioned above the vast majority of methods of solving nonlinear problems is based on *local* linearization. This means that the nonlinear equations are linearized about a vector running over a bounded subset B. This is true in case of topological methods: the Schauder fixed-point theorem and its generalizations, and the Leray-Schauder degree and its generalizations. But this is also true in case of iterative methods: the Banach contraction principle, the Newton-Kantorovič method, methods of contractors and contractor directions (Altman, 1977; 1980). Another important example is Kato's (1975) theory of quasilinear evolution equations and its generalization to nonreflexive Banach spaces (Altman, 1981).

However, there are some nonlinear problems involving "loss of derivatives" that are intractable, and there is no use of the above methods. In 1956 Nash introduced a procedure of attacking such problems that was based on a one-parameter family of smoothing operators. Moser (1966) introduced a scheme based on a modification of the Newton-Kantorovič method. The most important idea is Moser's degree of approximate linearization. Another modification of the Nash method was given by Hörmander (1976).

These methods can be characterized as iterative methods based on *global* linearization where B is nonbounded in contrast to *local* linearization mentioned above. In order to obtain smooth approximate solutions of the linearized equations, Nash used his smoothing operators, while Moser used his degree of approximate linearization.

The new idea behind the Nash-Moser technique was the motivation in our attempt to solve the problem of existence of solutions to nonlinear evolution equations. At this point, it should be emphasized that the Nash-Moser technique is designed for nonlinear operator equations and the methods mentioned above are not applicable to nonlinear evolution equations.

In Altman (1984a) a general theory of global linearization iterative methods (GLIM) is proposed for solving nonlinear operator equations:

1. GLIM-I is an iterative method with small steps, which is actually an iterative method of contractor directions.
2. GLIM-II is an iterative method with step size equal to one, which is independent of the notion of contractor directions.
3. GLIM-III is a rapidly convergent iteration method, based on a further development of Moser's essential technique.

Three new concepts are involved in the above methods: convex approximate linearization, which is similar to convex linearization implicitly employed in Altman (1981, 1983b); the order of approximate linearization, which generalizes Moser's degree of approximate linearization; and the degree of elliptic regularization. The best results can be obtained by making use of smoothing operators combined with elliptic regularization. GLIM-I seems to be the most universal and flexible method, but the degree of elliptic regularization has to be smaller than one. GLIM-III is not as flexible, but one is admissible as the degree of elliptic regularization, and also Moser's degree of approximate linearization should be greater than one.

Finally, let us mention that Definition 1.2 of approximate linearization (see Remark 1.1, Chapter 1) can also be applied to nonlinear evolution equations by using the same argument as in the case of nonlinear operator equations.

1. A GENERAL EXISTENCE THEOREM VIA GLIM-I

Let $Z \subset Y \subset X$ be Banach spaces with norms $\|\cdot\|_Z \geq \|\cdot\|_Y \geq \|\cdot\|_X$.

(A_0) We assume that there exist positive constants C, $\bar{s}$ with $0 < \bar{s} < 1$ such that

$$\|x\|_Y \leq C\|x\|_X^{1-\bar{s}} \cdot \|x\|_Z^{\bar{s}} \tag{1.0}$$

Given $0 < b$, denote by $C(0,b;X)$ the Banach space of all continuous functions $x = x(t)$ defined on the interval $[0,b]$ with values in X and the norm

$$\|x\|_{\infty,X} = \sup_t [\|x(t)\|_X: \quad 0 \leq t \leq b]$$

In the same way the norms $\|y\|_{\infty,Y}$ and $\|z\|_{\infty,Z}$ are defined for Y and Z. Denote by $C^1(0,b;X)$ the vector space of all continuously differentiable functions from [0,b] to X. Let W_0 be an open ball in Y with center x_0 in Z and radius r_0. Put $V_0 = W_0 \cap Z$ and let V_1 be the closure of V_0 in Y.

Let $F\colon [0,b] \times V_1 \to X$ be a nonlinear mapping and consider the Cauchy problem

$$Px(t) \equiv \frac{dx}{dt} + F(t,x) + f(t,x) = 0 \qquad 0 \le t \le b,\ x(0) = x_0 \quad (1.1)$$

where $f\colon [0,b] \times V_1 \to X$ is also a nonlinear mapping.

Let G be the set of functions

$$x \in C(0,b;V_0(\|\cdot\|_Z)) \cap C^1(0,b;X)$$

with $x(0) = x_0$ and $\|x - x_0\|_{\infty,Y} < r_0$.

We assume that the mapping F is differentiable in the following sense. For each $(t,x) \in [0,b] \times G$, there exists a linear operator $F'(t,x)$ such that $\varepsilon^{-1}\|F(\cdot, x + \varepsilon h) - F(\cdot,x) - \varepsilon F(\cdot,x)h\|_{\infty,X} \to 0$ as $\varepsilon \to 0+$, where $h \in C(0,b;Z) \cap C^1(0,b;X)$.

We make the following assumptions.

(A_1) Let $\{x_n\} \subset G$ be a Cauchy sequence in C(0,b;Y) and let

$$\{h_n\} \subset C(0,b;Y) \cap C^1(0,b;X)$$

be bounded in C(0,b;Y). Then $\varepsilon_n \to 0$ implies

$$\varepsilon_n^{-1}\|F(\cdot, x_n + \varepsilon_n h_n) - F(\cdot,x_n) - \varepsilon_n F'(\cdot,x_n)h_n\|_{\infty,X} \to 0$$

as $n \to \infty$. The functions F,f are continuous in the following sense:

$$\|x_n - x\|_{\infty,Y} \to 0 \qquad \text{implies} \qquad \|F(\cdot,x_n) - F(\cdot,x)\|_{\infty,X} \to 0$$

as $n \to \infty$, and the same is true for f.

There exists a constant q_0 such that

$$\|f(\cdot, x + \varepsilon h) - f(\cdot,x)\|_{\infty,X} \le q_0\varepsilon\|h\|_{\infty,X} \quad (1.2)$$

(A_2) There exists a constant $C_0 > 0$ with the following property. For $x \in G$ and g with $\|g\|_{\infty,X} < \infty$, if h is a solution of the equation

$$\frac{dh}{dt} + F'(t,x)h + g = 0 \qquad 0 \le t \le b,\ h(0) = 0 \tag{1.3}$$

then

$$\|h\|_{\infty,X} \le bC_0\|g\|_{\infty,X} \tag{1.4}$$

(A_3) For $x \in G$, the linearized equation

$$\frac{dz}{dt} + F'(t,x)z + F(t,x) - F'(t,x)x + f(t,x) = 0 \tag{1.5}$$

$0 \le t \le b$, $z(0) = 0$, admits smooth approximate solutions of order (μ,ν,σ) with $0 \le \eta < 1$ in the sense of the following definition.

1.1 DEFINITION Let $\mu > 0$, $\nu \ge 0$, $\sigma \ge 0$ be given numbers. Then the linearized equation (1.5) admits smooth approximate solutions of order (μ,ν,σ) if there exists a constant $M > 0$ which has the following property. For every $x \in G$, $K > 1$, and $Q > 1$, if $\|x\|_{\infty,Z} <$ K then there exist a residual (error) function y and a function z such that

$$\|z\|_{\infty,Z} \le MQK^{\nu} \tag{1.6}$$

$$\|y\|_{\infty,X} \le MQ^{-\mu}K^{\sigma} \tag{1.7}$$

and

$$\frac{dz}{dt} + F'(t,x)z + F(t,x) - F'(t,x)x + f(t,x) + y = 0 \qquad 0 \le t \le b,\ z(0) = 0 \tag{1.8}$$

Now for $x \in G$ let z be a solution of the equation (1.8) and put $z = x + h$. Then obviously h is a solution of the equation

$$\frac{dh}{dt} + F'(t,x)h + Px + y = 0 \qquad 0 \le t \le b,\ h(0) = 0 \tag{1.9}$$

and we get

$$\|h\|_{\infty,X} \le bC_0(\|Px\|_{\infty,X} + MQ^{-\mu}K^{\sigma}) \tag{1.10}$$

and

$$\|h\|_{\infty,Z} \le \|x\|_{\infty,Z} + MQK^{\nu} \tag{1.11}$$

1.1 LEMMA Let $x \in G$ and

$$q_0 bC_0 = \bar{q} < \frac{q}{2} < \frac{1}{2} \tag{1.12}$$

If h is a solution of the equation (1.9), then there exist $0 < \varepsilon \le 1$ such that

$$\|P(x + \varepsilon h) - (1 - \varepsilon)Px\|_{\infty,X} \le \varepsilon q\|Px\|_{\infty,X} \tag{1.13}$$

and

$$\|h\|_{\infty,X} \le bC_0(1 + \bar{q})\|Px\|_{\infty,X} \tag{1.14}$$

Proof: We use the following identity:

$$\begin{aligned} P(x + \varepsilon h) - (1 - \varepsilon)Px = {} & [F(\cdot, x + \varepsilon h) - F(\cdot,x) - \varepsilon F'(\cdot,x)h] \\ & + \varepsilon[dh/dt + F'(t,x)h + Px + y] - \varepsilon y \\ & + [f(\cdot, x + \varepsilon h) - f(\cdot,x)] \end{aligned}$$

Hence, we obtain by (1.2),

$$\begin{aligned} \|P(x + \varepsilon h) - (1 - \varepsilon)Px\|_{\infty,X} \le {} & \|F(\cdot, x + \varepsilon h) - F(\cdot,x) \\ & - \varepsilon F'(\cdot,x)h\|_{\infty,X} + \varepsilon\|y\|_{\infty,X} \\ & + q_0\varepsilon\|h\|_{\infty,X} \end{aligned}$$

and if $Q > 1$ is such that

$$\|y\|_{\infty,X} \le MQ^{-\mu}K^{\sigma} \le 2^{-1}\bar{q}\|Px\|_{\infty,X} \tag{1.15}$$

then we obtain

$$\begin{aligned} & \|P(x + \varepsilon h) - (1 - \varepsilon)Px\|_{\infty,X} \\ & \le \|F(\cdot, x + \varepsilon h) - F(\cdot,x) - \varepsilon F'(\cdot,x)h\|_{\infty,X} \\ & \quad + \varepsilon(q_0 bC_0 + \bar{q})\|Px\|_{\infty,X} \le \varepsilon q\|Px\|_{\infty,X} \end{aligned} \tag{1.15a}$$

whence relation (1.13) follows if

$$\|F(\cdot, x + \varepsilon h) - F(\cdot,x) - \varepsilon F'(\cdot,x)h\|_{\infty,X} \le \varepsilon(q - 2\bar{q})\|Px\|_{\infty,X} \quad \blacksquare$$

1.2 LEMMA Let $x_n \in G$ and

$$\|x_n\|_{\infty,Z} < A \exp(\alpha(1 - q)t_n) = K_n \tag{1.16}$$

where $t_n > 0$, and

$$\mu(1 - \nu) - \sigma > 0 \tag{1.17}$$

where $\alpha > 1$ and A are such that

$$\alpha(1 - q) - 1 > 0 \quad \text{and} \quad \alpha > [\mu(1 - \nu) - \sigma]^{-1} \tag{1.18}$$

and

$$M(2M)^{1/\mu}(\bar{q}p_0)^{-1/\mu} < A^{1-\nu-\sigma/\mu}[\alpha(1 - q) - 1] \tag{1.19}$$

where $p_0 = \|Px_0\|_{\infty,X}$. Then there exists a number $Q = Q_n$ such that

$$2MQ^{-\mu}K_n^{\sigma} < \bar{q}p_0 \exp(-(1 - q)t_n) \tag{1.20}$$

and

$$\|x_n + \varepsilon h_n\|_{\infty,Z} < A \exp(\alpha(1 - q)(t_n + \varepsilon)) \tag{1.21}$$

where h_n is a solution of the equation (1.9) with $x = x_n$, i.e.,

$$\frac{dh_n}{dt} + F'(t,x_n)h_n + Px_n + y_n = 0 \qquad 0 \le t \le b,\ h_n(0) = 0 \tag{1.22}$$

which satisfies relation (1.11) for $x = x_n$, $Q = Q_n$, and $K = K_n$.

Proof: Relation (1.11) for $x = x_n$ implies

$$\begin{aligned}\|x_n + \varepsilon h_n\|_{\infty,Z} &\le \|x_n\|_{\infty,Z} + \varepsilon\|h_n\|_{\infty,Z} \\ &\le (1 + \varepsilon)\|x_n\|_{\infty,Z} + \varepsilon MQ_nK^{\nu}\end{aligned}$$

Hence, (1.20) follows if

$$(1 + \varepsilon)A \exp(\alpha(1 - q)t_n) + \varepsilon MQK_n^{\nu} < A \exp(\alpha(1 - q)(t_n + \varepsilon)) \tag{1.23}$$

But it is easily seen that

$$\alpha(1 - q) - 1 = \min_{0\le\varepsilon\le 1} [\exp(\alpha(1 - q)\varepsilon) - (1 + \varepsilon)]/\varepsilon \tag{1.24}$$

Hence, by (1.15), $Q = Q_n$ is subject to

$$Q < (\alpha(1 - q) - 1)A^{1-\nu}M^{-1} \exp(\alpha(1 - \nu)(1 - q)t_n) \tag{1.25}$$

On the other hand, (1.20) holds if

$$(2M)^{1/\mu}(\bar{q}p_0)^{-1/\mu}A^{\sigma/\mu} \exp((\sigma\alpha + 1)(1 - q)t_n/\mu) < Q \tag{1.26}$$

Hence, relations (1.25) and (1.26) are satisfied if conditions (1.17) to (1.19) hold. This completes the proof of the lemma. ∎

We shall now construct an iterative method of contractor directions as follows. Put $x_0(t) \equiv x_0$, and assume that $x_0, x_1, \ldots, x_n \in G$ are known and satisfy the following relations for all indices $i \le n$,

$$\|x_i\|_{\infty,Z} < A \exp(\alpha(1 - q)t_i) = K_i \tag{1.27}$$

and

$$\|Px_i\|_{\infty,X} \le \|Px_0\|_{\infty,X} \exp(-(1 - q)t_i) \tag{1.28}$$

where α and A are subject to (1.18) and (1.19). Next let z_n be a solution of the equation (1.8) with $x = x_n$, $y = y_n$, find $Q = Q_n$ from Lemma 1.2, and put $z_n = x_n + h_n$ so that h_n becomes a solution of the equation (1.22). Now with $0 < \varepsilon_n \le 1$ to be determined, put

$$x_{n+1} = x_n + \varepsilon_n h_n \qquad t_{n+1} = t_n + \varepsilon_n \qquad t_0 = 0 \tag{1.29}$$

that is,

$$x_{n+1} = (1 - \varepsilon_n)x_n + \varepsilon_n z_n$$

which justifies the term "convex approximate linearization." It follows from Lemma 1.2 that relation (1.27) holds for x_{n+1} and $t_{n+1} = t_n + \varepsilon_n$. Also relation (1.28) holds for x_{n+1} and $t_{n+1} = t_n + \varepsilon_n$. This results from relation (1.13) with $x = x_n$, $h = h_n$, and $\varepsilon = \varepsilon_n$, in which the right-hand term $\|Px_n\|_{\infty,X}$ is replaced by its estimate. In fact, we have

$$\begin{aligned}\|Px_{n+1}\|_{\infty,X} &\le (1 - (1 - q)\varepsilon_n)\|Px_n\|_{\infty,X} \\ &\le \exp(-(1 - q)\varepsilon_n)\|Px_n\|_{\infty,X} \\ &\le \|Px_0\|_{\infty,X} \exp(-(1 - q)(t_n + \varepsilon_n))\end{aligned}$$

by induction. In order to determine ε_n, let c be such that $2\bar{q}/q < c < 1$, and put

$$\Phi(\varepsilon,h,x) = \varepsilon^{-1}\|P(x + \varepsilon h) - (1 - \varepsilon)Px\|_{\infty,X}$$

If

$$\Phi(1,h_n,x_n) \le q\|Px_0\|_{\infty,X} \exp(-(1 - q)t_n)$$

then put $\varepsilon_n = 1$ in (1.29). If

$$\Phi(1,h_n,x_n) > q\|Px_0\|_{\infty,X} \exp(-(1 - q)t_n)$$

then there exists $0 < \varepsilon < 1$ such that

$$\begin{aligned}cq\|Px_0\|_{\infty,X} \exp(-(1 - q)t_n) &\le \Phi(\varepsilon,h_n,x_n) \\ &\le q\|Px_0\|_{\infty,X} \exp(-(1 - q)t_n)\end{aligned} \tag{1.30}$$

and put $\varepsilon_n = \varepsilon$ in (1.29).

We are now in a position to prove the following.

1.1 THEOREM In addition to the hypotheses (A_0) to (A_3), suppose that conditions (1.17) to (1.19) are satisfied, and b' is such that

$$\begin{aligned}N = [(1 - q)\delta]^{-1} \exp((1 - q)\delta)C[b'C_0(1 + \bar{q})\|Px_0\|_{\infty,X}]^{1-\bar{s}} \\ \times [\alpha(1 - q)A]^{\bar{s}} < r_0\end{aligned} \tag{1.31}$$

where $\bar{s}$ is such that

$$\delta = 1 - (1 + \alpha)\bar{s} > 0 \qquad \alpha < \frac{1 - \bar{s}}{\bar{s}} \tag{1.32}$$

Then equation (1.1) with b replaced by b' has a solution x, and

$$\|x_n - x\|_{\infty,Y} \to 0 \qquad \text{as } n \to \infty \tag{1.33}$$

where $\{x_n\}$ is determined by (1.29), and

$$\|x_n - x_0\|_{\infty,Y} < r_0 \qquad \text{for all } n \tag{1.34}$$

Proof: First prove that the sequence $\{h_n\}$ is bounded in $C(0,b;Y)$. In fact, we have by virtue of (1.11) with $x = x_n$, (1.16) and (1.11)

$$\begin{aligned}\|h_n\|_{\infty,Z} &\le \|x_n\|_{\infty,Z} + MQ_n K_n^{\nu} \\ &< A\exp(\alpha(1 - q)t_n) + (\alpha(1 - q) - 1)A\exp(\alpha(1 - q)t_n)\end{aligned}$$

Hence, we get

$$\|h_n\|_{\infty,Z} \le \alpha(1 - q)A\exp(\alpha(1 - q)t_n) \tag{1.35}$$

It results from (1.10) with $x = x_n$ and (1.28) that

$$\begin{aligned}\|h_n\|_{\infty,X} &\le bC_0(\|Px_n\|_{\infty,X} + MQ_n^{-\mu}K_n^{\sigma}) \\ &\le bC_0(p_0\exp(-(1 - q)t_n) + \bar{q}p_0\exp(-(1 - q)t_n)\end{aligned}$$

Hence,

$$\|h_n\|_{\infty,X} \le bC_0(1 + \bar{q})\|Px_0\|_{\infty,X}\exp(-(1 - q)t_n) \tag{1.36}$$

Relations (1.35), (1.36), and (A_0) imply

$$\begin{aligned}\|h_n\|_{\infty,Y} &\le C\|h_n\|_{\infty,X}^{1-\bar{s}}\|h_n\|_{\infty,Z}^{\bar{s}} \\ &\le C[bC_0(1 + \bar{q})\|Px_0\|_{\infty,X}]^{1-\bar{s}} \\ &\quad \times [\alpha(1 - q)A]^{\bar{s}}\exp(-(1 - q)\delta t_n)\end{aligned} \tag{1.37}$$

with δ as in (1.32). We have

$$\begin{aligned}\sum_{n=0}^{\infty}\varepsilon_n\exp(-(1 - q)\delta t_n) &= \sum_{n=0}^{\infty}(t_{n+1} - t_n)\exp(-(1 - q)\delta t_n \\ &= \sum_{n=0}^{\infty}(t_{n+1} - t_n)\exp(-(1 - q)\delta t_{n+1}) \\ &\quad \times \exp((1 - q)\delta\varepsilon_n)\end{aligned}$$

$$\le \exp((1-q)\delta) \sum_{n=0}^{\infty} \int_{t_n}^{t_{n+1}} \exp(-(1-q)\delta t)\, dt$$

$$= \exp((1-q)\delta) \int_0 \exp(-(1-q)\delta t)\, dt$$

$$= \exp((1-q)\delta)[(1-q)\delta]^{-1}$$

Hence,

$$\sum_{n=0}^{\infty} \varepsilon_n \|h_n\|_{\infty,Y} \le N \tag{1.38}$$

where N is given by (1.31). Finally, we prove that

$$\|Px_n\|_{\infty,X} \to 0 \quad \text{as } n \to \infty \tag{1.39}$$

Since $t_0 = 0$ and $t_n = \Sigma_{i=0}^{n-1} \varepsilon_i$, we consider two cases: (a) the sequence $\{\varepsilon_n\}$ does not converge to 0; then $t_n \to \infty$, and (1.39) follows from (1.28); and (b) $\varepsilon_n \to 0$ as $n \to \infty$. In this case we have by virtue of (1.15a) with $x = x_n$, $\varepsilon = \varepsilon_n$, and $h = h_n$ that

$$\Phi(\varepsilon_n, h_n, x_n) \le \varepsilon_n^{-1} \|F(\cdot, x_n + \varepsilon_n h_n) - F(\cdot, x_n) - \varepsilon_n F'(\cdot, x_n) h_n\|_{\infty,X}$$
$$+ 2\bar{q} \|Px_0\|_{\infty,X} \exp(-(1-q)t_n)$$

by virtue of (1.15a) and (1.28).

Hence, we get from the first inequality (1.30)

$$(cq - 2\bar{q}) \|Px_0\|_{\infty,X} \exp(-(1-q)t_n)$$
$$\le \varepsilon_n^{-1} \|F(\cdot, x_n + \varepsilon_n F'(\cdot, x_n) h_n\|_{\infty,X}$$

But the right-hand side of the last inequality converges to 0, by virtue of (A_1), since $\{h_n\}$ is bounded and $\{x_n\}$ is convergent in C(0,b;Y) by (1.37) and (1.38). Hence, relation (1.39) follows from (1.28), since $\exp(-(1-q)t_n) \to 0$ as $n \to \infty$. This completes the proof of the theorem. ∎

The idea behind the convex approximate linearization is that equation (1.9) becomes a consequence of (1.8) where the norm of the term $F(t,x) - F'(t,x)x + f(t,x) + y$ can be controlled, which is not the case in (1.9), since Px contains the derivative dx/dt, by (1.1). The significance of this fact is shown in the next chapter, where smoothing operators are involved.

Another advantage appears in the following case. Suppose that F in the equation (1.1) is replaced by $F = A + \bar{F}$, where A is a linear operator. Then we have

$$F(t,x) - F'(t,x)x = \bar{F}(t,x) - \bar{F}'(t,x)x$$

that is, in the linearized equation (1.5), the term $F(t,x) - F'(t,x)x$ can be replaced by $\bar{F}(t,x) - \bar{F}'(t,x)x$, which does not contain A.

2. A GENERAL EXISTENCE THEOREM VIA GLIM-II

Let $Z \subset Y \subset X$ be Banach spaces with norms $\|\cdot\|_Z \geq \|\cdot\|_Y \geq \|\cdot\|_X$.

We assume that there exist positive constants $C, \bar{s}$ with $0 < \bar{s} < 1$ such that

$$(A_0) \quad \|x\|_Y \leq C\|x\|_X^{1-\bar{s}}\|x\|_Z^{\bar{s}}$$

Given $0 < b$, denote by $C(0,b;X)$ the Banach space of all continuous functions $x = x(t)$ defined on the interval $[0,b]$ with values in X and the norm

$$\|x\|_{\infty,X} = \sup_t [\|x(t)\|: \quad 0 \leq t \leq b]$$

In the same way the norms $\|y\|_{\infty,Y}$ and $\|z\|_{\infty,Z}$ are defined for Y and Z, respectively. Denote by $C^1(0,b;X)$ the vector space of all continuously differentiable functions from $[0,b]$ to X.

Let W_0 be an open ball in Y with center x_0 in Z and radius $r > 0$. Put $V_0 = W_0 \cap Z$ and let V_1 be the $\|\cdot\|_Y$-closure of V_0.

Let F: $[0,b] \times V_1 \to X$ be a nonlinear mapping, and consider the Cauchy problem

$$Px(t) \equiv \frac{dx}{dt} + F(t,x) = 0 \qquad 0 \leq t \leq b, \; x(0) = x_0 \tag{2.1}$$

Let G be the set of functions

$$x \in C(0,b,V_0(\|\cdot\|_Z)) \cap C^1(0,b;X) \qquad \|x - x_0\|_{\infty,Y} < r$$

with $x(0) = x_0$. We assume that the mapping F is differentiable in the following sense. For each $(t,x) \in [0,b] \times G$, there exists a linear operator $F'(t,x)$ such that

$$\varepsilon^{-1}\|F(\cdot, x + \varepsilon h) - F(\cdot,x) - \varepsilon F'(\cdot,x)\|_{\infty,X} \to 0$$

as $\varepsilon \to 0+$, where $h \in C(0,b;Z) \cap C^1(0,b;X)$.

2.1 DEFINITION Let μ, ν, and σ be positive constants with $\nu < 1$. Then the linearized equation

$$\frac{dz}{dt} + F'(t,x)z + F(t,x) - F'(t,x)x = 0 \qquad (2.2)$$

$$0 \le t \le b, \ x(0) = x_0$$

admits smooth approximate solutions of order (μ,ν,σ) if there exists a constant $M > 0$ which has the following property. For every $x \in G$, $K > 1$, and $Q > 1$, if $\|x\|_{\infty,Z} < K$, then there exist a residual (error) function y and a function z such that

$$\|z\|_{\infty,Z} \le MQK^{\nu} \qquad 0 < \nu < 1 \qquad (2.3)$$

$$\|y\|_{\infty,X} \le MQ^{-\mu}K^{\sigma} \qquad \sigma,\mu > 0 \qquad (2.4)$$

and

$$\frac{dz}{dt} + F'(t,x)z + F(t,x) - F'(t,x)x + y = 0 \qquad (2.5)$$

$$0 \le t \le b, \ z(0) = 0$$

We make the following assumptions.

(A_1) There exists a constant $L > 0$ such that

$$\|F(t,u) - F(t,v) - F'(t,u)(u - v)\|_X \le L\|u - v\|_X\|u - v\|_Y \qquad (2.6)$$

for all $t \in [0,b]$, $u,v \in V_0$.

The function F is continuous in the following sense:

$$\|x_n - x\|_{\infty,Y} \to 0 \quad \text{implies} \quad \|F(\cdot,x_n) - F(\cdot,x)\|_{\infty,X} \to 0$$

as $n \to \infty$.

(A_2) There exists a constant $C > 0$ with the following property. For $x \in G$ with $\|g\|_{\infty,X} < \infty$, if h is a solution of the equation

$$\frac{dh}{dt} + F'(t,x)h + g = 0 \qquad 0 \le t \le b,\ h(0) = 0 \tag{2.6}$$

then

$$\|h\|_{\infty,X} \le bC\|g\|_{\infty,X} \tag{2.7}$$

(A_3) For $x \in G$, the linearized equation (2.2) admits approximate smooth solutions in the sense of Definition 2.1.

Now for $x \in G$ let z be a solution of the equation (2.5), and put $z = x + h$. Then obviously h is a solution of the equation

$$\frac{dh}{dt} + F'(t,x)h + Px + y = 0 \qquad 0 \le t \le b,\ h(0) = 0 \tag{2.8}$$

and we get by virtue of (2.4), (2.3), and (2.7) with g replaced by $Px + y$,

$$\|h\|_{\infty,X} \le bC(\|Px\|_{\infty,X} + MQ^{-\mu}K^{\sigma}) \tag{2.9}$$

and

$$\|h\|_{\infty,Z} \le \|x\|_{\infty,Z} + MQK^{\nu} \tag{2.10}$$

2.1 LEMMA Let $x \in G$ with $\|x - x_0\|_{\infty,Y} < r$, $0 < \bar{q} < q < 1$ be arbitrary fixed numbers, and let h be a solution of the equation (1.8) with $\|h\|_{\infty,Y} < r$. If $K > 1$, $\|x\|_{\infty,Z} < K$, and $Q > 1$ is such that

$$\|y\|_{\infty,X} \le MQ^{-\mu}K^{\sigma} \le \bar{q}\|Px\|_{\infty,X} \tag{2.11}$$

then

$$\|P(x + h)\|_{\infty,X} \le q\|Px\|_{\infty,X} \tag{2.12}$$

and

$$\|h\|_{\infty,X} \le bC(1 + \bar{q})\|Px\|_{\infty,X} \tag{2.13}$$

provided that b satisfies the relation

$$2LrbC \le q - \bar{q} \tag{2.14}$$

Proof: We use the following identity:

$$P(x + h) = [F(\cdot, x + h) - F(\cdot,x) - F'(\cdot,x)h + [dh/dt + F'(t,x)h + Px + y] - y$$

Hence, we obtain by (2.8)

$$\begin{aligned}\|P(x + h)\|_{\infty,X} &\le L\|h\|_{\infty,X}\|h\|_{\infty,Y} + \|y\|_{\infty,X} \\ &\le Lr\|h\|_{\infty,X} + \bar{q}\|Px\|_{\infty,X} \\ &\le 2LrbC\|Px\|_{\infty,X} + \bar{q}\|Px\|_{\infty,X} \\ &\le (q - \bar{q})\|Px\|_{\infty,X} + \bar{q}\|Px\|_{\infty,X}\end{aligned}$$

Hence, relation (2.12) results. Relation (2.13) follows from (2.9) and (2.11). ■

2.2 LEMMA Put $x_0(t) \equiv x_0$, and with $\|x_n - x_0\|_{\infty,Y} < r$ let

$$\|x_n\|_{\infty,Z} < Aq^{-\alpha n} = K_n \tag{2.15}$$

$$\|Px_n\|_{\infty,X} \le p_0 q^n \tag{2.16}$$

where $p_0 = \|Px_0\|_{\infty,X}$, and α, A are chosen so as to satisfy the relations

$$\alpha[\mu(1 - \nu) - \sigma] > 1 \tag{2.17}$$

$$M^{1+1/\mu}(\bar{q}p_0)^{-1/\mu}(q^{-\alpha} - 2)^{-1} < A^{1-\nu-\sigma/\mu} \qquad q < 2^{-1/\alpha} \tag{2.18}$$

Then there exists a number $Q = Q_n$ such that

$$MQ_n^{-\mu}K_n^{\alpha} \le \bar{q}p_0q^n \tag{2.19}$$

and

$$\|x_n + h_n\|_{\infty,Z} < Aq^{-\alpha(n+1)} = K_{n+1} \tag{2.20}$$

where h_n is a solution of the equation (2.8) with $x = x_n$, i.e.,

$$\frac{dh_n}{dt} + F'(t,x_n)h_n + Px_n + y_n = 0 \qquad 0 \le t \le b,\ h_n(0) = 0 \tag{2.21}$$

which satisfies relations (2.9) and (2.10) for $x = x_n$.

Proof: Relation (2.10) for $x = x_n$ implies

$$\|x_n + h_n\|_{\infty,Z} \le \|x_n\|_{\infty,Z} + \|h_n\|_{\infty,Z} \le 2\|x_n\|_{\infty,Z} + MQ_nK_n^{\nu}$$

since $h_n = z_n - x_n$. Hence (2.20) follows if

$$2Aq^{-\alpha n} + MQ_nK_n^{\nu} < Aq^{-\alpha(n+1)} \tag{2.22}$$

or

$$Q_n < M^{-1}A^{1-\nu}q^{-\alpha(1-\nu)n}(q^{-\alpha} - 2) \tag{2.22a}$$

On the other hand, (2.19) holds if

$$M^{1/\mu}(\bar{q}p_0)^{-1/\mu}A^{\sigma/\mu}q^{-(\alpha\sigma+1)n/\mu} < Q_n \tag{2.23}$$

Hence, relations (2.22a) and (2.23) are satisfied if conditions (2.17) and (2.18) hold. ■

We shall now construct the following iterative process. Put $x_0(t) \equiv x_0$, and assume that $x_0, x_1, \ldots, x_n$ are known and satisfy the relations (2.15) and (2.16) of Lemma 2.2.

Then put

$$x_{n+1} = z_n = x_n + h_n \tag{2.24}$$

where z_n and h_n are solutions of the equation (2.5) with $x = x_n$, $y = y_n$, and (2.21), respectively, i.e., z_n is an approximate solution of the linearized equation (2.2). It results from (2.20) that relation (2.15) holds true for $n + 1$, and relation (2.12) with $x = x_n$, $h = h_n$ implies that condition (2.16) is also satisfied for $n + 1$.

2.3 LEMMA The following estimate holds with N, β as in (1.28).

$$\sum_{n=0}^{\infty} \|h_n\|_{\infty,Y} \le \frac{N}{1 - q^{\beta}} \tag{2.25}$$

Proof: We have by (2.13) with $x = x_n$, $h = h_n$, and (2.16),

$$\|h_n\|_{\infty,X} \le bC(1 + \bar{q})p_0 q^n \tag{2.26}$$

It results from (2.10) with $x = x_n$, $h = h_n$, (2.15), and (2.22) that

$$\begin{aligned} \|h_n\|_{\infty,Z} &\le \|x_n\|_{\infty,Z} + MQ_n K_n^{\nu} \\ &\le Aq^{-\alpha n} + Aq^{-\alpha(n+1)} - 2Aq^{-\alpha n} \\ &= A(q^{-\alpha} - 1)q^{-\alpha n} \end{aligned}$$

Hence, with $q < 2^{-1/\alpha}$ we get

$$\|h_n\|_{\infty,Z} \le A(q^{-\alpha} - 1)q^{-\alpha n} \tag{2.27}$$

Relations (2.26), (2.27), and (A_0) imply

$$\|h_n\|_{\infty,Y} \le C\|h_n\|_{\infty,X}^{1-\bar{s}}\|h_n\|_{\infty,Z}^{\bar{s}} \le Nq^{\beta n} \tag{2.28}$$

where

$$N = C[bC(1 + \bar{q})p_0]^{1-\bar{s}}[A(q^{-\alpha} - 1)]^{\bar{s}}$$

and $\beta = [1 - \bar{s}(1 + \alpha)] > 0$, if α is chosen so that $\alpha < (1 - \bar{s})/\bar{s}$ and $q < 2^{-1/\alpha}$, and in addition, conditions (2.17) and (2.18) are satisfied. ∎

We are now in a position to prove the following.

2.1 THEOREM In addition to the assumptions (A_0) to (A_3), suppose that N and β are determined by (2.28); $\alpha < (1 - \bar{s})/\bar{s}$ is chosen so as to satisfy (2.17), $0 < \bar{q} < q < 2^{-1/\alpha}$, and A is chosen so as to satisfy (2.18). Let $0 \le b' \le b$ be such that relations (2.14) and

$$\frac{N}{1 - q^{\beta}} < r \tag{2.29}$$

hold. Then equation (2.1) with b replaced by b' has a solution x, and

$$\|x_n - x\|_{\infty,Y} \to 0 \qquad \text{as } n \to \infty \tag{2.30}$$

where $\{x_n\} \subset G$ is determined by (2.24), and

$$\|x_n - x_0\|_{\infty,Y} < r \qquad \text{for all } n \tag{2.31}$$

Proof: The convergence (2.30) results from (2.25) and (2.24), and so does relation (2.31), by (2.29). Since $\|Px_n\|_{\infty,X} \to 0$ as $n \to \infty$, by virtue of (2.16), it follows from the continuity of F (see A_1) that $Px = 0$. This completes the proof of the theorem. ∎

2.1 REMARK The equation (2.1) can be replaced by the more general equation (1.1) with f satisfying condition (1.2), where $\varepsilon = 1$. The above reasoning carries over with minor changes.

3. A GENERAL EXISTENCE THEOREM VIA GLIM-III

An iterative method is presented for solving general nonlinear evolution equations in nonreflexive Banach spaces. The method utilizes and further develops some of the essential techniques due to Moser (1966).

Also Moser's definition of approximate solutions of the linearized equations is being generalized so as to enhance the scope of applications of the method. The crucial idea that makes the method work is the concept of "convex approximate linearization," which is similar to the "convex linearization" implicitly employed in Altman (1981). It should be emphasized that Moser's method is designed for nonlinear operator equations and is not applicable to nonlinear evolution equations.

Let $Z \subset Y \subset X$ be Banach spaces with norms $\|x\|_Z \geq \|x\|_Y \geq \|x\|_X$. Let $C > 0$ and $0 < \bar{s} < 1$ be such that

(A_0) $\quad \|x\|_Y \le C\|x\|_X^{1-\bar{s}}\|x\|_Z^{\bar{s}}$

Given $0 < b$, denote by $C(0,b;X)$ the Banach space of all continuous X-valued functions $x = x(t)$ with norm

$$\|x\|_{\infty,X} = \sup_t [\|x(t)\|_X : \quad 0 \le t \le b]$$

Similar notation is used for Y and Z. Denote by $C^1(0,b;X)$ the vector space of all continuously differentiable functions from $[0,b]$ into X.

Let F: $[0,b] \times Y \to X$ be a nonlinear mapping. Consider the Cauchy problem

$$Px(t) \equiv \frac{dx}{dt} + F(t,x) = 0 \qquad 0 \le t \le b, \; x(0) = x_0 \tag{3.1}$$

Let W_0 be an open ball with center $x_0 \in Z$ and radius $r > 0$, $V_0 = W_0 \cap Z$, V_1 being the $\|\cdot\|_Y$-closure of V_0. Let G be the set of functions $x \in C(0,b;V_0(\|\cdot\|_Z)) \cap C^1(0,b;X)$ with $x(0) = x_0$, $\|x - x_0\|_{\infty,Y} < r$.

Assumptions:

(A_1) $F(t,x)$ is continuous in the sense:

$$x_n \in G \text{ and } \|x_n - x\|_{\infty,Y} \to 0 \text{ imply } \|F(\cdot,x_n) - F(\cdot,x)\|_{\infty,X} \to 0$$

as $n \to \infty$. The linear derivative operator, $F'(t,x)$, exists, i.e., for $x \in G$, $\varepsilon^{-1}\|F(\cdot, x + \varepsilon h) - F(\cdot,x) - \varepsilon F'(\cdot,x)h\|_{\infty,X} \to 0$ as $\varepsilon \to 0+$, where $h \in C(0,b;Z) \cap C^1(0,b;X)$.

(A_2) F' satisfies

$$\|F(\cdot, x + h) - F(\cdot,x) - F'(\cdot,x)h\|_{\infty,X} \le M\|h\|_{\infty,X}^{2-\beta}\|h\|_{\infty,Z}^{\beta} \tag{3.2}$$

for some $M > 0$, $0 \le \beta < 1$, and all $x \in G$.

(A_3) There is a constant $C > 0$ with the following property. If $\|g\|_{\infty,X} < \infty$ and h is a solution of the equation

$$\frac{dh}{dt} + F'(t,x)h + g = 0 \qquad 0 \le t \le b, \; h(0) = 0 \tag{3.3}$$

then

$$\|h\|_{\infty,X} \le C\|g\|_{\infty,X} \tag{3.4}$$

The linearized equations

$$\frac{dz}{dt} + F'(t,x)z + F(t,x) - F'(t,x)x = 0 \qquad (3.4)$$
$$0 \le t \le b, \ x(0) = x_0$$

admit approximate solutions in the following sense.

3.1 DEFINITION Let μ, ν, and σ be positive constants. Then the linearized equation admits approximate solutions of order (μ,ν,σ) if there exists a constant $M > 0$ with the following property. For every $x \in G$, $K > 1$, and $Q > 1$ if $\|x\|_{\infty,Z} < K$, then there exist a residual (error) function y and a function z such that

$$\|z\|_{\infty,Z} \le MQK^{\nu} \tag{3.6}$$

$$\|y\|_{\infty,X} \le MQ^{-\mu}K^{\sigma} \tag{3.7}$$

and

$$\frac{dz}{dt} + F'(t,x)z + F(t,x) - F'(t,x)x + y = 0 \tag{3.8}$$
$$0 \le t \le b, \ z(0) = x_0$$

Now put $z = x + h$, $x \in G$. Then obviously h is a solution of the equation

$$\frac{dh}{dt} + F'(t,x)h + Px + y = 0 \qquad 0 \le t \le b, \ h(0) = 0 \tag{3.9}$$

The following method generalizes some of Moser's (1966) technique, which he applied to nonlinear operator equations. This method is also applicable to nonlinear operator equations. Moreover, since one assumes $\nu = \sigma = 1 + \alpha$ $(0 < \alpha < 1)$, instead of $\nu = \sigma = 1$ in Moser's case, the scope of applications of this method is enhanced.

Let $K_{n+1} = K_n^{\tau+\alpha}$ $(1 < \tau < 2)$, and put

$$x_0(t) \equiv x_0 \qquad x_{n+1} = x_n + h_n \qquad x_n \in G \tag{3.10}$$

where z_n is a solution of (3.8), and Q_n from (3.6) and (3.7) is chosen so as to satisfy the following induction assumptions,

$$\|x_n\|_{\infty,Z} < K_n \qquad x_n \in G \tag{3.11}$$

$$\|Px_n\|_{\infty,X} \le K_n^{-\lambda} \tag{3.12}$$

for some $\lambda > 0$ to be determined.

(A_4) Let $\nu = \sigma = 1 + \alpha$ $(0 < \alpha < 1)$, $\mu > 1$, and let τ, $\lambda < \mu$ be such that

$$(\mu - \lambda)^{-1}(1 + \alpha(1 + \lambda) + \mu) < \tau < 2 - \alpha_0 < 2 - \alpha \tag{3.13}$$

$$0 < 2\lambda < \frac{\mu(1 - \alpha_0) - (1 + \alpha_0)}{1 - \alpha_0} \tag{3.14}$$

where $0 < \alpha < \alpha_0 < 1$ is such that (see Chapter 1)

$$\mu > \frac{1 + \alpha_0}{1 - \alpha_0} > 1 \tag{3.15}$$

and let β be such that

$$0 < \beta < \mu\lambda(\alpha_0 - \alpha)[(1 + \alpha)(1 + \mu) + \lambda(2 + \mu)]^{-1} < 1 \tag{3.16}$$

3.1 REMARK One can put $\alpha_0 = (\mu - 1)/(\mu + 3)$ in (3.13) to (3.16).

3.1 THEOREM Suppose that assumptions (A_0) to (A_4) are satisfied. Then there exists a constant $K_0(M,\beta,\mu,\lambda) > 1$ such that

$$\|F(\cdot,x_0)\|_{\infty,X} < K_0^{-\lambda} \qquad \|x_0\|_Z < K_0$$

$$C(2C)^{1-\bar{s}} \sum_{n=0}^{\infty} K_n^{-\delta} < r \tag{3.17}$$

where $\delta = (1 - \bar{s})\lambda - \bar{s}\tau > 0$ provided

$$\bar{s} < \frac{\lambda}{\lambda + 2} \tag{3.18}$$

and $K_{n+1} = K_n^{\tau+\alpha}$ with τ as in (3.13), imply that equation (3.1) has a solution x which is a limit of $\{x_n\}$ and

$$\|x_n - x\|_{\infty,Y} \to 0 \qquad \text{as } n \to \infty \tag{3.19}$$

where $\{x_n\}$ is determined by (3.10).

Proof: To verify the induction assumption (3.11) we use (3.10) and (3.6) with $z = z_n$, $K = K_n$, $\nu = 1 + \alpha$, and choose Q so as to satisfy

$$2MQK_n^{1+\alpha} < K_{n+1} = K_n^{\tau+\alpha} \tag{3.20}$$

Then obviously

$$\|x_{n+1}\|_{\infty,Z} = \|z_n\|_{\infty,Z} < MQK_n^{1+\alpha} < K_{n+1} \tag{3.21}$$

To verify the induction assumption (3.12) we use relation (3.2) with $x = x_n$, $h = h_n$, where h_n satisfies (3.9) with $y = y_n$, and the following identity,

$$\begin{aligned} Px_{n+1} &= P(x_n + h_n) \\ &= [F(\cdot, x_n + h_n) - F(\cdot,x_n) - F'(\cdot,x_n)h_n] \\ &\quad + [dh_n/dt + F'(t,x_n)h_n + Px_n + y_n] - y_n \end{aligned}$$

Hence, by virtue of (3.22) and (3.24), we get

$$\begin{aligned} \|Px_{n+1}\|_{\infty,X} &\le M\|h_n\|_{\infty,X}^{2-\beta}\|h_n\|_{\infty,Z}^{\beta} + \|y_n\|_{\infty,X} \\ &\le M(2C)^{2-\beta}K_n^{-\lambda(2-\beta)}(2M)^{\beta}(QK_n^{1+\alpha})^{\beta} + MQ^{-\mu}K_n^{1+\alpha} \\ &\le c[K_n^{-\lambda(2-\beta)}(QK_n^{1+\alpha})^{\beta} + Q^{-\mu}K_n^{1+\alpha}] \end{aligned}$$

where $c > M$ is some constant, since

$$\begin{aligned} \|h_n\|_{\infty,X} &\le C(\|Px\|_{\infty,X} + \|y_n\|_{\infty,X}) \\ &\le C(K_n^{-\lambda} + MQ^{-\mu}K_n^{1+\alpha}) \end{aligned}$$

Hence,

$$\|h_n\|_{\infty,X} \le 2CK_n^{-\lambda} \tag{3.22}$$

provided that Q is chosen so as to satisfy

$$MQ^{-\mu}K_n^{1+\alpha} < K_n^{-\lambda} \tag{3.23}$$

Since $z_n = x_n + h_n$, we get

$$\|h_n\|_{\infty,Z} \le \|x_n\|_{\infty,Z} + \|z_n\|_{\infty,Z}$$
$$\le K_n + MQK_n^{1+\alpha}$$

Hence,

$$\|h_n\|_{\infty,Z} < 2MQK_n^{1+\alpha} < K_{n+1} \tag{3.24}$$

by virtue of (3.20). Thus (3.12) can be verified if Q can be chosen so as to satisfy

$$c[K_n^{-\lambda(2-\beta)}(QK_n^{1+\alpha})^{\beta} + Q^{-\mu}K_n^{1+\alpha}] < K_{n+1}^{-\lambda} \tag{3.25}$$

Therefore, by virtue of (3.20), (3.23), and (3.25), it is sufficient to show that the following inequalities can be satisfied for some $Q > 1$, and $K_{n+1} = K_n^{\tau+\alpha}$.

$$cK_n^{1+\alpha}Q < K_{n+1} \tag{3.26}$$

$$c(K_n^{1+\alpha}Q)^{\beta}K_n^{-\lambda(2-\beta)} < K_{n+1}^{-\lambda} \tag{3.27}$$

$$cK_n^{1+\alpha}Q^{-\mu} < K_{n+1}^{-\lambda} \tag{3.28}$$

For the proof see Section 3, Chapter 1.

Finally, we get from (A_0), (3.22), and (3.24),

$$\|h_n\|_{\infty,Y} \le C\|h_n\|_{\infty,X}^{1-\bar{s}}\|h_n\|_{\infty,Z}^{\bar{s}}$$
$$\le C(2CK_n^{-\lambda})^{1-\bar{s}}K_n^{\tau\bar{s}} = C(2C)^{1-\bar{s}}K_n^{-\delta}$$

where $\delta > 0$ is determined by (3.17). Hence,

$$\sum_{n=0}^{\infty} \|g_b\|_{\infty,Y} \le C(2C)^{1-\bar{s}} \sum_{n=0}^{\infty} K_n^{-\delta} < r$$

It follows from (3.29) that $\|x_n\|_{\infty,Y} < r$, and (3.19) holds for some x, which is a solution of equation (3.1), by virtue of (3.12) and (A_1). It is clear that relation (3.17) is satisfied if K_0 is sufficiently large. ∎

Consider now the case in Definition 3.1 where $\nu = \sigma = 1$. Here, instead of (A_4) we make the following assumption.

(A'_4) Suppose that

$$0 < \lambda + 1 < \frac{\mu + 1}{2}$$

and

$$0 < \beta < \frac{\lambda}{\lambda + 1} \frac{\mu}{\mu + 1} \left(1 - 2\frac{\lambda + 1}{\mu + 1}\right)$$

where β is from (3.2). Then τ is a number satisfying the following:

$$1 < \left(1 - \frac{\lambda + 1}{\mu + 1}\right)^{-1} < \tau < 2$$

3.2 THEOREM Theorem 3.1 remains valid if (A_4) is replaced by (A'_4), $\nu = \sigma = 1$, and $K_{n+1}^{\tau} = K_n$, $n = 0, 1, \ldots$.

Proof: The proof is exactly the same as that of Theorem 3.1 provided that the inequalities (3.26) to (3.28) are replaced by the following ones.

$$cK_nQ < K_{n+1} = K_n^{\tau} \tag{3.30}$$

$$c(K_nQ)^{\beta}K_n^{-\lambda(2-\beta)} < K_{n+1}^{-\lambda} \tag{3.31}$$

$$cK_nQ^{-\mu} < K_{n+1}^{-} \tag{3.32}$$

Moser (1966) has shown that $Q > 1$ satisfying (3.30) to (3.32) can be found. For the proof see Section 3, Chapter 1. ∎

4. AN APPLICATION OF THE THEORY OF C_0-SEMIGROUPS

Sufficient conditions for the existence of a constant in (A_2) can be obtained by making use of the theory of C_0-semigroups.

Denote by $G(X)$ the set of all negative infinitesimal generators of C_0-semigroups $\{U(t)\}$ on X. If $-A \in G(X)$, then $\{U(t)\} = \{e^{-tA}\}$, $0 \le t < \infty$, is the semigroup generated by $-A$. For $x \in G$, put $A(t) = A(t,x(t)) = F'(t,x(t))$. We assume that A is a function from $[0,b] \times V_0$ into $G(X)$, and $A(t)$ is stable in X (see Kato, 1970; 1973; 1975; Yosida, 1968; Tanabe, 1979). We also assume that the evolution operator generated by $A(t,x(t))$ exists and satisfies the following condition,

$$\|U(t,s;x)\|_{X\to X} \le M_X \tag{4.1}$$

for all $x \in G$ and some constant $M_X > 0$. Let us mention that the evolution operators $U(t,x;x)$ can be replaced by their approximations $U_m(t,s;x)$ generated by the appropriate step functions (see Altman, 1982b). Condition (4.1) implies the existence of a constant in (A_2).

4.1 REMARK In addition to the assumptions made above, suppose that

$$\|F(t,u) - F(t,v) - F'(t,u)(u - v)\|_X \le k\|u - v\|_X\|u - v\|_Y$$

for all $t \in [0,b]$, $u,v \in V_1$, and some $k = k(F)$. Then the solution to the equation (3.1) is unique.

Proof: By (4.1), the evolution operator $U(t,s;x)$ generated by $A(t,x) = F'(t,x)$ exists in X, and

$$\frac{d}{ds} U(t,s;x)v = U(t,s;x)A(t,x)v \qquad v \in Y,\ 0 \le s \le t \le b$$

Now let $x(\cdot)$ and $y(\cdot)$ be two different solutions to (3.1) with $m = \sup_t[\|x(t) - y(t)\|_X;\ 0 \le t \le b]$. Then we get

$$x(t) - y(t) = [U(t,s;x\cdot)(x(s) - y(s)]_{s=0}^{s=t}$$

$$x(t) - y(t) = \int_0^\tau \frac{\partial}{\partial s} [U(t,s;x)(x(s) - y(s))] \, ds$$

$$= \int_0^t U(t,s;x)[x'(s) - y'(s) + A(t,x)(x(s) - y(s))] \, ds$$

$$= \int_0^t U(t,s;x)[F(t,x) - F(t,y) - A(t,x)(x,s) - y(s))] \, ds$$

which implies that

$$\|x(t) - y(t)\|_X \le 2M_X kr \int_0^t \|x(s) - y(s)\|_X \, ds$$

Hence, we obtain

$$m \le 2M_X rbm$$

which implies that $m = 0$ if $2M_X rbm < 1$. ∎

For a more general uniqueness theorem, see Graff (1979).

8

Smoothing Operators Combined With Elliptic Regularization and the Degree of Elliptic Regularization for Nonlinear Evolution Equations

This chapter shows how to construct approximate solutions of the linearized evolution equations. As in Chapter 2 we use smoothing operators to accomplish this goal. The best results can be obtained by a combination of smoothing operators and elliptic regularization. In this way no "gain of derivatives" is needed for the estimate of the solution of the linearized equation. Elliptic regularization can also be used without smoothing operators. But then an appropriate gain of derivatives is needed for the estimate of the solution of the linearized equation. As in Chapter 2 the degree of elliptic regularization is also introduced. Having constructed the approximate solutions of the linearized equations, one then makes use of the results of Chapter 7 in order to prove their convergence to a solution of the nonlinear evolution equation. All three methods (GLIM) presented in Chapter 7 can now be applied successfully.

1. A GENERAL EXISTENCE THEOREM FOR NONLINEAR EVOLUTION EQUATIONS VIA GLIM-I IN A SCALE OF BANACH SPACES

Let $\{X_j\}$ with $0 \le j \le p$ be a scale of Banach spaces with increasing norms such that $i < j$ implies $X_j \subset X_i$ and $\|\cdot\|_j \ge \|\cdot\|_i$ and let $0 < m_1 < m_2 < \bar{p} < p$.

(A_0) We assume that there exists a one-parameter family of linear smoothing operators S_θ, $\theta \ge 1$ (see Nash, 1956; Moser, 1966) such that

$$\|(I - S_\theta)x\|_0 \le C\theta^{-m_1}\|x\|_{m_1}$$

$$\|(I - S_\theta)x\|_{m_1} \le C\theta^{-(m_2-m_1)}\|x\|_{m_2}$$

$$\|S_\theta x\|_p \le C\theta^{p-m_2}\|x\|_{m_2}$$

for some constant $C > 0$, where I is the identity mapping. We also assume that

$$\|x\|_j \le C\|x\|_r^{1-\lambda}\|x\|_i^\lambda$$

for $j = (1 - \lambda)r + \lambda i;\ 0 \le \lambda \le 1$.

Given $0 < b$, denote by $C(0,b;X_j)$ the Banach space of all continuous functions $x = x(t)$ defined on the interval $[0,b]$ with values in X_j and norm

$$\|x\|_{\infty,j} = \sup_t[\|x(t)\|_j:\quad 0 \le t \le b]$$

for $j = 0,\ m_1,\ m_2,\ \bar{p},\ p$ and s to be determined. Let $W_0 \subset X_s$ be an open ball with center $x_0 \in X_p$ and radius $r_0 > 0$. Put $V_0 = W_0 \cap X_p$ and let V_s be the $\|\cdot\|_s$-closure of V_0.

Let $F,\ f\colon\ [0,b] \times X_s \to X_0$ be two nonlinear mappings and consider the Cauchy problem

$$Px(t) \equiv \frac{dx}{dt} + F(t,x) + f(t,x) = 0 \qquad (1.1)$$
$$0 \le t \le b,\ x(0) = X_0$$

Using the same notation as in Section 1, let G be the set of functions

$$x \in C(0,b;V_0(\|\cdot\|_p)) \cap C^1(0,b;X_0)$$

with $x(0) = x_0$ and $\|x - x_0\|_{\infty,s} < r_0$.

(A_1) We assume that F is continuous in the following sense:

$$\|x_n - x\|_{\infty,s} \to 0 \quad \text{implies} \quad \|F(\cdot,x_n) - F(\cdot,x)\|_{\infty,0} \to 0 \qquad (1.2)$$

as $n \to \infty$, and so is f. We also assume that F is differentiable in the following sense. For each $(t,x) \in [0,b] \times G$, there exists a linear operator $F'(t,x)$ such that

$$\varepsilon^{-1}\|F(\cdot, x + \varepsilon h) - F(\cdot,x) - \varepsilon F'(\cdot,x)h\|_{\infty,0} \to 0 \tag{1.3}$$

as $\varepsilon \to 0+$, where $h \in C(0,b;X_p) \cap C'(0,b;X_0)$.

There exists a constant q_0 such that

$$\|f(t, x + \varepsilon h) - f(t,x)\|_0 \le q_0 \in \|h\|_0 \tag{1.4}$$

for all $(t,x) \in [0,b] \times V_0$, $h \in X_s$.

There exists a constant $C > 0$ such that

$$\|F'(t,x)h\|_{m_1} \le C\|h\|_{m_2} \tag{1.5}$$

and

$$\|F'(t,x)h\|_0 \le C\|h\|_{m_1} \tag{1.6}$$

for all $(t,x) \in [0,b] \times V_0$.

There exists a constant $C > 0$ such that

$$\|F(t,x) - F'(t,x)x + f(t,x)\|_{m_2} \le C\|x\|_{\bar{p}} \tag{1.7}$$

is satisfied for all $(t,x) \in [0,b] \times V_0$.

(A_2) Let $\{x_n\} \subset G$ be a Cauchy sequence in $C(0,b;X_s)$ and let $\{h_n\}$ be bounded in $C(0,b;X_s)$; then $\varepsilon_n \to 0+$ implies that

$$\varepsilon_n^{-1}\|F(\cdot, x_n + \varepsilon_n h_n) - F(\cdot,x_n) - \varepsilon_n F'(\cdot,x_n)h_n\|_{\infty,0} \to 0 \tag{1.8}$$

(A_3) There exists a linear (regularizing) operator $L = L(\eta)$ such that

$$\|Lz\|_0 < C\|z\|_{m_2} \tag{1.9}$$

for some constant $C > 0$, and the modified linearized equation

$$\frac{dz}{dt} + F'(t,x)z + \eta Lz + F(t,x) - F'(t,x)x + f(t,x) = 0 \tag{1.10}$$
$$0 \le t \le b, \quad z(0) = 0$$

with small $0 < |\eta| < 1$ and $(t,x) \in [0,b] \times G$ has a solution $\bar{z}$ such that

$$\|\bar{z}(t)\|_{m_2} \le C|\eta|^{-\bar{k}}\|F(t,x) - F'(t,x)x(t) - f(t,x)\|_{m_2} \tag{1.11}$$

for some $\bar{k} \ge 0$ to be determined and $C > 0$.

The number $\bar{k} \ge 0$ is called the degree of elliptic regularization.

(A_4) For $(t,x) \in [0,b] \times G$, if h is a solution of the equation

$$\frac{dh}{dt} + F'(t,x)h + g = 0 \qquad 0 \le t \le b,\ h(0) = 0 \tag{1.12}$$

then

$$\|h\|_{\infty,0} \le C_0\|g\|_{\infty,0} \tag{1.13}$$

for some constant $C_0 > 0$.

In order to simplify the notation we assume that $0 < \eta < 1$.

For $(t,x) \in [0,b] \times G$, let $\bar{z}$ be a solution of the equation (1.10) satisfying (1.11). Then the following estimates hold.

1.1 LEMMA The following relations hold for $(t,x) \in [0,b] \times G$ with $\|x\|_{\infty,p} < K$, $K > 1$; $0 < \eta < 1$.

$$\|(I - S_\theta)\frac{d\bar{z}}{dt}\|_{\infty,0} \le M_1(\theta^{-m_1} + \eta)\eta^{-k}K^{\bar{\nu}} \tag{1.14}$$

$$\|F'(\cdot,x)(I - S_\theta)\bar{z}\|_{\infty,0} \le M_2\theta^{-(m_2-m_1)}\eta^{-\bar{k}}K^{\bar{\nu}} \tag{1.15}$$

$$\|z\|_{\infty,p} \equiv \|S_\theta\bar{z}\|_{\infty,p} \le M_3\theta^{p-m_2}\eta^{-\bar{k}}K^{\bar{\nu}} \tag{1.16}$$

for some $M_1,M_2,M_3 > 0$ and $z = S_\theta\bar{z}$, where $\bar{\nu} = (\bar{p} - \rho)/(p - \rho)$ with ρ to be determined and $0 < \bar{k} < 1$.

Proof: We get from (A_0), (1.5), and (1.7),

$$\begin{aligned}
&\|(I - S_\theta)\bar{z}(t)\|_0 \\
&= \|(I - S_\theta)[F'(t,x)\bar{z}(t) + \eta L\bar{z}(t) + F(t,x) - F'(t,x)x + f(t,x)]\|_0 \\
&\le C\theta^{-m_1}[\|F'(t,x)\bar{z}(t)\|_{m_1} + \|F(t,x) - F'(t,x)x + f(t,x)\|_{m_1}] + C\eta\|L\bar{z}(t)\|_0 \\
&\le C\theta^{-m_1}[\|\bar{z}(t)\|_{m_2} + \|x(t)\|_{\bar{p}}] + C\eta\|\bar{z}(t)\|_{m_2}
\end{aligned}$$

Hence, relation (1.14) follows, since

$$\|x(t)\|_{\bar{p}} \le C\|x(t)\|_{\rho}^{1-\bar{\nu}}\|x(t)\|_{p}^{\bar{\nu}} \le Cr^{1-\bar{\nu}}K^{\bar{\nu}}$$

and

$$\|\bar{z}(t)\|_{m_2} \le C_1\eta^{-k}\|x(t)\|_{\bar{p}} \le C_2\eta^{-k}K^{\bar{\nu}} \tag{1.17}$$

by (1.11) and (1.7). We have

$$\begin{aligned}\|F'(t,x)(I - S_\theta)\bar{z}(t)\|_0 &\le C\|(I - S_\theta)\bar{z}(t)\|_{m_1} \\ &\le C\theta^{-(m_2-m_1)}\|\bar{z}(t)\|_{m_2}\end{aligned}$$

and (1.15) follows from (1.17). We get by virtue of (A_0),

$$\|S_\theta\bar{z}(t)\|_p \le C\theta^{p-m_2}\|\bar{z}(t)\|_{m_2}$$

and (1.16) results from (1.17). ∎

Consider the linearized equation

$$\frac{dz}{dt} + F'(t,x)z + F(t,x) - F'(t,x)x + f(t,x) = 0 \qquad 0 \le t \le b,\ z(0) = 0 \tag{1.18}$$

and put $z = S_\theta\bar{z}$, where $\bar{z}$ is a solution of the modified linearized equation (1.10) satisfying (1.11).

1.2 LEMMA The following estimate holds,

$$\left\|\frac{dz}{dt} + F'(\cdot,x)z + F(\cdot,x) - F'(\cdot,x)x + f(\cdot,x)\right\|_{\infty,0} \le M_4(\theta^{-m} + \eta)\eta^{-\bar{k}}K^{\bar{\nu}} \tag{1.19}$$

for $x \in G$ with $\|x\|_{\infty,p} < K$, $K > 1$, and some $M_4 > 0$, where

$$m = \min(m_1,\ m_2 - m_1) \tag{1.20}$$

Proof: We have

$$\|\frac{dz}{dt} + F'(t,x)z(t) + F(t,x) - F'(t,x)x + f(t,x)\|_0$$
$$\leq \|\frac{d\bar{z}}{dt} + F'(t,x)\bar{z}(t) + \eta L\bar{z}(t) + F(t,x)x + f(t,x)\|_0$$
$$+ \|(I - S_\theta)\frac{d\bar{z}}{dt}\|_0 + \|F'(t,x)(I - S_\theta)\bar{z}\|_0 + \eta\|L\bar{z}(t)\|_0$$

and (1.19) follows, by virtue of (1.10) with $z = \bar{z}$, and (1.14), (1.15), and (1.20). Now suppose there exist ν,μ with

$$\bar{\nu} < \nu < 1 \quad \text{and} \quad \mu(1 - \nu) - \nu > 0 \tag{1.21}$$

such that for $Q > 1$, $K > 1$, one can find $0 < \eta < 1$ and $\theta > 1$ to satisfy

$$\theta^{p-m_2}\eta^{-\bar{k}}K^{\bar{\nu}} < QK^{\nu} \tag{1.22}$$

$$(\theta^{-m} + \eta)\eta^{-\bar{k}}K^{\bar{\nu}} < Q^{-\mu}K^{\nu} \tag{1.23}$$

Then relation (1.19) can be written as

$$\|\frac{dz}{dt} + F'(\cdot,x)z + F(\cdot,x) - F'(\cdot,x)\|_{\infty,0} \leq 2M_4Q^{-\mu}K^{\nu} \tag{1.24}$$

by (1.21) to (1.23). ∎

1.1 DEFINITION Let $\mu > 0$, $\nu \geq 0$, $\sigma \geq 0$ be given numbers. Then the linearized equation (1.18) admits approximate solutions of order (μ,ν,σ) if there exists a constant $M > 0$ with the following property. For every $x \in G$, $K > 1$, and $Q > 1$, if $\|x\|_{\infty,p} < K$, then there exist a residual (error) function y and a function z such that

$$\|z\|_{\infty,p} \leq MQK^{\nu} \tag{1.25}$$

$$\|y\|_{\infty,0} \leq MQ^{-\mu}K^{\sigma} \tag{1.26}$$

and

$$\frac{dz}{dt} + F'(t,x)z + F(t,x) - F'(t,x)x + f(t,x) + y = 0 \tag{1.27}$$
$$0 \leq t \leq b, \quad z(0) = 0$$

1.3 LEMMA The linearized equation (1.18) admits approximate solutions of order (μ,ν,σ) in the sense of Definition 1.1, with $\sigma = \nu < 1$

to be determined from (1.23) and (1.22), where

$$\bar{\nu} = \frac{\bar{p} - \rho}{p - \rho} \tag{1.28}$$

with ρ as in the proof of Lemma 1.6, Chapter 2.

Proof: The proof results from (1.22) to (1.24). ∎

Thus we have shown that if $\bar{z}$ is a solution of the equation (1.10) satisfying (1.11), then $z = S_\theta \bar{z}$ with appropriate choice of θ is an approximate solution of order (μ,ν,σ) (with $\sigma = \nu$) of the linearized equation (1.18); that is, there exist a constant $M > 0$ and a function z which satisfy relations (1.25) to (1.27).

Now put $z = x + h$. Then it follows from (1.27) that h is a solution of the equation

$$\frac{dh}{dt} + F'(t,x)h + Px + y = 0 \qquad 0 \le t \le b,\ h(0) = 0 \tag{1.29}$$

and we get from (1.25), (1.26), and (1.13),

$$\|h\|_{\infty,0} \le C_0(\|Px\|_{\infty,0} + MQ^{-\mu}K^{\sigma}) \tag{1.30}$$

$$\|h\|_{\infty,p} \le \|x\|_{\infty,p} + MQK^{\nu} \tag{1.31}$$

where $\sigma = \nu$.

1.4 LEMMA Let $x \in G$ with $\|x\|_{\infty,p} < K$, $K > 1$, and let

$$q_0 b C_0 = \bar{q} < \frac{q}{2} < \frac{1}{2} \tag{1.32}$$

If h is a solution of the equation (1.29), then there exists $0 < \varepsilon \le 1$ such that

$$\|P(x + \varepsilon h) - (1 - \varepsilon)Px\|_{\infty,0} \le \varepsilon q\|Px\|_{\infty,0} \tag{1.33}$$

and

$$\|h\|_{\infty,0} \le C_0(1 + \bar{q})\|Px\|_{\infty,0}$$

Proof: The proof is exactly the same as in Lemma 1.1, Chapter 7. ∎

1.5 LEMMA For $x_n \in G$, let

$$\|x_n\|_{\infty,p} < A \exp(\alpha(1 - q)t_n) = K_n \tag{1.35}$$

where $t_n > 0$, and let α, A, $\bar{q}$, and q be such that

$$\max([\mu(1 - \nu) - \nu]^{-1}, (1 - q)^{-1}) < \alpha \tag{1.36}$$

and

$$\mu(1 - \nu) - \nu > 0 \tag{1.37}$$

and

$$M(2M)^{1/\mu}(\bar{q}p_0)^{-1/\mu} < A^{1-\nu-\nu/\mu}[\alpha(1 - q) - 1] \qquad \alpha > 1 \tag{1.38}$$

where $p_0 = \|Px_0\|_{\infty,0}$ and $\bar{q}, q_0$ satisfy (1.32). Then there exists a number $Q = Q_n$ such that

$$2MQ_n^{-\mu}K_n^{\nu} < \bar{q}p_0 \exp(-(1 - q)t_n) \tag{1.39}$$

and

$$\|x_n + \varepsilon h_n\|_{\infty,p} < A \exp(\alpha(1 - q)(t_n + \varepsilon)) \tag{1.40}$$

for all $0 < \varepsilon \leq 1$, where h_n is a solution of the equation (1.29) with $x = x_n$, $y = y_n$, provided that relations (1.30) and (1.31) hold with $x = x_n$, $h = h_n$, $K = K_n$, and $Q = Q_n$.

Proof: The proof of Lemma 1.2, Chapter 7, carries over with $\|\cdot\|_X$, $\|\cdot\|_Z$ replaced by $\|\cdot\|_0$, $\|\cdot\|_p$, respectively. ∎

Relations (1.22) and (1.23) impose an additional condition on Q, which results from the proof of the following key lemma for the degree $\bar{k}$ of elliptic regularization.

1.6 LEMMA There exist ν with $\bar{\nu} < \nu < 1$, and μ, which satisfies (1.21) and such that for $K > 1$, $Q > 1$, one can find $0 < \eta < 1$ and $\theta > 1$ which satisfy (1.22) and (1.23), provided

$$0 < \bar{k} < 1 \tag{1.40a}$$

$$\bar{\nu} = \frac{\bar{p} - \rho}{p - \rho} \tag{1.40b}$$

with ρ as in the proof of Lemma 1.6, Chapter 2.

Proof: For the proof see Lemma 1.6, Chapter 2. ∎

Notice that m_2 in (1.22) and (1.9) can be replaced by $\bar{p}$.

Thus ν and μ can be found to satisfy (1.20) to (1.22) and consequently both conditions (1.25) and (1.26), Chapter 7, can be satisfied.

In the same way as in Chapter 7, we construct the following method of contract or directions:

$$x_{n+1} = x_n + \varepsilon_n h_n \qquad t_{n+1} = t_n + \varepsilon_n \qquad t_0 = 0 \tag{1.41}$$

where h_n is a solution of the equation (1.29) with $x = x_n$, $y = y_n$, and induction assumptions

$$\|x_n\|_{\infty,p} < A \exp(\alpha(1 - q)t_n) = K_n \tag{1.42}$$

$$\|Px_n\|_{\infty,0} \le \|Px_0\|_{\infty,0} \exp(-(1 - q)t_n) \tag{1.43}$$

where α, A, and q are subject to

$$\max([\mu(1 - \nu) - \nu]^{-1}, (1 - q)^{-1}) < \alpha < \frac{p - s}{s} \tag{1.44}$$

and (1.38) so that $Q = Q_n$ can be found to satisfy (1.25), (1.26), Chapter 7. The choice of $\{\varepsilon_n\}$ is exactly the same as in Chapter 7 with $\|\cdot\|_X$ replaced by $\|\cdot\|_0$. The estimate (1.37) is still valid with $\|\cdot\|_Y$ replaced by $\|\cdot\|_s$, where s is chosen so as to satisfy

$$\max([\mu(1 - \nu) - \nu]^{-1}, (1 - q)^{-1}) < \frac{p - s}{s} \tag{1.45}$$

which is a consequence of (1.32) with $\bar{s} = s/p$, Chapter 7.

1.1 THEOREM In addition to assumptions (A_0) to (A_4) suppose that conditions (1.44) above, and (1.31), Chapter 7, are satisfied. Then the equation (1.1) has a solution x such that

$$\|x_n - x\|_{\infty,s} \to 0 \qquad \text{as } n \to \infty$$

where $\{x_n\} \subset G$ is determined by (2.41) and $\|x\|_{\infty,s} < r_0$.

Proof: The proof of Theorem 1.1, Chapter 7, carries over. ∎

1.1 REMARK The advantage of convex approximate linearization is made more obvious by the application of smoothing operators. In fact, suppose that, following the standard routine, we linearize (approximately) the equation (1.1) and obtain the equation (1.29) instead of (1.27). Then it is impossible to obtain an estimate for the solution of (1.29) in terms which are free of the norm of the derivative dx/dt as in the case of (1.27) [see (1.11)].

2. A GENERAL EXISTENCE THEOREM FOR NONLINEAR EVOLUTION EQUATIONS VIA GLIM-II IN A SCALE OF BANACH SPACES

Let $\{X_j\}$ with $0 \le j \le p$ be a scale of Banach spaces with increasing norms such that $i < j$ implies $X_j \subset X_i$ and $\|\cdot\|_j \ge \|\cdot\|_i$ and let $0 < m_1 < m_2 < \bar{p} < p$.

(A_0) We assume that there exists a one-parameter family of linear smoothing operators S_θ, $\theta \ge 1$, which satisfy the same conditions as in Section 1.

Given $0 < b$, denote by $C(0,b;X_j)$ the Banach space of all continuous functions $x = x(t)$ defined on the interval $(0,b)$ with values in X_j and norm

$$\|x\|_{\infty,j} = \sup_t [\|x(t)\|_j: \quad 0 \le t \le b]$$

for $j = 0, m_1, m_2, \bar{p}, p$ and s to be determined. Let $W_0 \subset X_s$ be an open ball with center $x_0 \in X_p$ and radius $r_0 > 0$. Put $V_0 = W_0 \cap X_p$ and let V_s be the $\|\cdot\|_s$-closure of V_0.

Let $F, f\colon (0,b) \times X_s \to X_0$ be two nonlinear mappings and consider the Cauchy problem

$$Px(t) \equiv \frac{dx}{dt} + F(t,x) + f(t,x) = 0 \qquad 0 \le t \le b,\ x(0) = X_0 \quad (2.1)$$

Using the same notation as in Section 1, let G be the set of functions

$$x \in C(0,b;V_0(\|\cdot\|_p)) \cap C^1(0,b;X_0)$$

with $x(0) = x_0$ and $\|x - x_0\|_{\infty,s} < r_0$.

(A_1) We assume that F is continuous in the following sense:

$$\|x_n - x\|_{\infty,s} \to 0 \quad \text{implies} \quad \|F(\cdot,x_n) - F(\cdot,x)\|_{\infty,0} \to 0 \tag{2.2}$$

as $n \to \infty$, and so is f. We also assume that F is differentiable in the following sense. For each $(t,x) \in (0,b) \times G$, there exists a linear operator $F'(t,x)$ such that

$$\varepsilon^{-1}\|F(\cdot, x + \varepsilon h) - F(\cdot,x) - \varepsilon F'(\cdot,x)h\|_{\infty,0} \to 0 \tag{2.3}$$

as $\varepsilon \to 0+$, where $h \in C(0,b;X_p) \cap C'(0,b;X_0)$.

There exists a constant q_0 such that

$$\|f(t, x + h) - f(t,x)\|_0 \le q_0\|h\|_0 \tag{2.4}$$

for all $(t,x) \in (0,b) \times V_0$, $h \in X_s$.

There exists a constant $C > 0$ such that

$$\|F'(t,x)h\|_{m_1} \le C\|h\|_{m_2} \tag{2.5}$$

and

$$\|F'(t,x)h\|_0 \le C\|h\|_{m_1} \tag{2.6}$$

for all $(t,x) \in (0,b) \times V_0$.

There exists a constant $C > 0$ such that

$$\|F(t,x) - F'(t,x)x + f(t,x)\|_{m_2} \le C\|x\|_{\bar{p}} \tag{2.7}$$

is satisfied for all $(t,x) \in (0,b) \times V_0$.

(A_2) There exists a constant $C > 0$ such that

$$\|F(t,u) - F(t,v) - F'(t,u)(u - v)\|_0 \le C\|u - v\|_0\|u - v\|_s \tag{2.8}$$

for all $t \in (0,b)$; $u,v \in V_0$.

(A_3) There exists a linear (regularizing) operator $L = L(\eta)$ such that

$$\|Lz\|_0 < C\|z\|_{m_2} \tag{2.9}$$

for some constant $C > 0$, and the modified linearized equation

$$\frac{dz}{dt} + F'(t,x)z + \eta Lz + F(t,x) - F'(t,x)x + f(t,x) = 0 \tag{2.10}$$
$$0 \le t \le b, \; z(0) = 0$$

with small $0 < |\eta| < 1$ and $(t,x) \in (0,b) \times G$ has a solution $\bar{z}$ such that

$$\|\bar{z}(t)\|_{m_2} \le C|\eta|^{-\bar{k}}\|F(t,x) - F'(t,x)x(t) - f(t,x)\|_{m_2} \tag{2.11}$$

for some $\bar{k} \ge 0$ to be determined and $C > 0$. The number $\bar{k} > 0$ is called the degree of elliptic regularization.

(A_4) For $(t,x) \in (0,b) \times G$ if h is a solution of the equation

$$\frac{dh}{dt} + F'(t,x)h + g = 0 \qquad 0 \le t \le b, \; h(0) = 0 \tag{2.12}$$

then

$$\|h\|_{\infty,0} \le C_0\|g\|_{\infty,0} \tag{2.13}$$

for some constant $C_0 > 0$.

In order to simplify the notation we assume that $0 < \eta < 1$.

For $(t,x) \in (0,b) \times G$, let $\bar{z}$ be a solution of the equation (2.10) satisfying (2.11). Then we have the estimates:

2.1 LEMMA The following relations hold for $(t,x) \in (0,b) \times G$ with $\|x\|_{\infty,p} < K$, $K > 1$; $0 < \eta < 1$.

$$\|(I - S_\theta)\frac{d\bar{z}}{dt}\|_{\infty,0} \le M_1(\theta^{-m_1} + \eta)\eta^{-\bar{k}}K^{\bar{\nu}} \tag{2.14}$$

$$\|F'(\cdot,x)(I - S_\theta)\bar{z}\|_{\infty,p} \le M_2\theta^{-(m_2-m_1)}\eta^{-\bar{k}}K^{\bar{\nu}} \tag{2.15}$$

$$\|z\|_{\infty,p} \equiv \|S_\theta\bar{z}\|_{\infty,p} \le M_3\theta^{p-m_2}\eta^{-\bar{k}}K^{\bar{\nu}} \tag{2.16}$$

for some $M_1,M_2,M_3 > 0$ and $z = S_\theta\bar{z}$, where $\bar{\nu} = (\bar{p} - \rho)/(p - \rho)$ with ρ to be determined and $0 < \bar{k} < 1$.

Proof: We get from (A_0), (2.5), and (2.7),

$$\|(I - S_\theta)\bar{z}(t)\|_0 = \|(I - S_\theta)[F'(t,x)\bar{z}(t) + \eta L\bar{z}(t) + F(t,x) - F'(t,x)x + f(t,x)]\|_0$$
$$\le C\theta^{-m_1}[\|F'(t,x)\bar{z}(t)\|_{m_1} + \|F(t,x) - F'(t,x)x + f(t,x)\|_{m_1}] + C\eta\|L\bar{z}(t)\|_0$$
$$\le C\theta^{-m_1}[\|\bar{z}(t)\|_{m_2} + \|x(t)\|_{\bar{p}}] + C\eta\|\bar{z}(t)\|_{m_2}$$

Hence, relation (2.14) follows, since

$$\|x(t)\|_{\bar{p}} \le C\|x(t)\|_\rho^{1-\bar{\nu}}\|x(t)\|_p^{\bar{\nu}} \le Cr^{1-\bar{\nu}}K^{\bar{\nu}}$$

and

$$\|\bar{z}(t)\|_{\bar{p}} \le C_1\eta^{-k}\|x(t)\|_{\bar{p}} \le C_2\eta^{-k}K^{\bar{\nu}} \tag{2.17}$$

by (2.11) and (2.7). We have

$$\|F'(t,x)(I - S_\theta)\bar{z}(t)\|_0 \le C\|(I - S_\theta)\bar{z}(t)\|_{m_1}$$
$$\le C\theta^{-(m_2-m_1)}\|\bar{z}(t)\|_{m_2},$$

and (2.15) follows from (2.17). We get by virtue of (A_0)

$$\|S_\theta\bar{z}(t)\|_p \le C\theta^{p-\bar{p}}\|\bar{z}(t)\|_{\bar{p}}$$

and (2.16) results from (2.17). ∎

Consider the linearized equation

$$\frac{dz}{dt} + F'(t,x)z + F(t,x) - F'(t,x)x + f(t,x) = 0 \tag{2.18}$$
$$0 \le t \le b, \quad z(0) = 0$$

and put $z = S_\theta\bar{z}$, where $\bar{z}$ is a solution of the modified linearized equation (2.10) satisfying (2.11).

2.2 LEMMA The following estimate holds.

$$\left\|\frac{dz}{dt} + F'(\cdot,x)z + F(\cdot,x) - F'(\cdot,x)x + f(\cdot,x)\right\|_{\infty,0}$$
$$\le M_4(\theta^{-m} + \eta)\eta^{-\bar{k}}K^{\bar{\nu}} \tag{2.19}$$

for $x \in G$ with $\|x\|_{\infty,p} < K$, $K > 1$, and some $M_4 > 0$, where

$$m = \min(m_1, m_2 - m_1) \tag{2.20}$$

Proof: We have

$$\left\|\frac{dz}{dt} + F'(t,x)z(t) + F(t,x) - F'(t,x)x + f(t,x)\right\|_0$$
$$\le \left\|\frac{d\bar{z}}{dt} + F'(t,x)\bar{z}(t) + \eta L\bar{z}(t) + F(t,x)x + f(t,x)\right\|_0$$
$$+ \left\|(I - S_\theta)\frac{d\bar{z}}{dt}\right\|_0 + \|F'(t,x)(I - S_\theta)\bar{z}\|_0 + \eta\|L\bar{z}(t)\|_0$$

and (2.19) follows, by virtue of (2.10) with $z = \bar{z}$, and (2.14), (2.15), and (2.20). Now suppose there exist ν,μ with

$$\bar{\nu} < \nu < 1 \qquad \text{and} \qquad \mu(1 - \nu) - \nu > 0 \tag{2.21}$$

such that for $Q > 1$, $K > 1$, one can find $0 < \eta < 1$ and $\theta > 1$ to satisfy

$$\theta^{p-m_2}\eta^{-\bar{k}}K^{\bar{\nu}} < QK^{\nu} \tag{2.22}$$
$$(\theta^{-m} + \eta)\eta^{-\bar{k}}K^{\bar{\nu}} < Q^{-\mu}K^{\nu} \tag{2.23}$$

Then relation (2.19) can be written as

$$\left\|\frac{dz}{dt} - F'(\cdot,x)z - F(\cdot,x) - F'(\cdot,x)x - f(\cdot,x)\right\|_{\infty,0}$$
$$\le 2M_4Q^{-\mu}K^{\nu} \tag{2.24}$$

by (2.21) to (2.23). ∎

2.1 DEFINITION Let $\mu > 0$, $\nu \ge 0$, $\sigma \ge 0$ be given numbers. Then the linearized equation (2.18) admits approximate solutions of order (μ,ν,σ) if there exists a constant $M > 0$ with the following property. For every $x \in G$, $K > 1$, and $Q > 1$, if $\|x\|_{\infty,p} < K$, then there exist a residual (error) function y and a function z such that

$$\|z\|_{\infty,p} \leq MQK^{\nu} \tag{2.25}$$

$$\|y\|_{\infty,0} \leq MQ^{-\mu}K^{\sigma} \tag{2.26}$$

and

$$\frac{dz}{dt} + F'(t,x)z + F(t,x) - F'(t,x)x + f(t,x) + y = 0 \qquad 0 \leq t \leq b,\ z(0) = x_0 \tag{2.27}$$

2.3 LEMMA The linearized equation (2.18) admits approximate solutions of order (μ,ν,σ) in the sense of Definition 2.1, with $\sigma = \nu < 1$ to be determined from (2.23) and (2.22).

Proof: The proof results from (2.22) to (2.24). ■

Thus we have shown that if $\bar{z}$ is a solution of the equation (2.10) satisfying (2.11), then $z = S_\theta \bar{z}$ with appropriate choice of θ is an approximate solution of order (μ,ν,σ), with $\sigma = \nu$, of the linearized equation (2.18); that is, there exist a constant $M > 0$ and a function z which satisfy relations (2.25) to (2.27).

Now put $z = x + h$. Then it follows from (2.27) that h is a solution of the equation

$$\frac{dh}{dt} + F'(t,x)h + Px + y = 0 \qquad 0 \leq t \leq b \qquad h(0) = 0 \tag{2.28}$$

and we get from (2.25), (2.26), and (2.13),

$$\|h\|_{\infty,0} \leq C_0(\|Px\|_{\infty,0} + MQ^{-\mu}K^{\sigma}) \tag{2.29}$$

$$\|h\|_{\infty,p} \leq \|x\|_{\infty,p} + MQK^{\nu} \tag{2.30}$$

where $\sigma = \nu$.

2.4 LEMMA Let $x \in G$ with $\|x\|_{\infty,p} < K$. Suppose that h is a solution of the equation (2.28). If $K > 1$ and $Q > 1$ are such that

$$\|y\|_{\infty,0} \leq MQ^{-\nu}K^{\nu} \leq \bar{q}\|Px\|_{\infty,0} \tag{2.31}$$

then

$$\|P(x + h)\|_{\infty,0} \leq q\|Px\|_{\infty,0} \tag{2.32}$$

$$\|h\|_{\infty,0} \le C_0(1 + \bar{q})\|Px\|_{\infty,0} \tag{2.33}$$

provided that

$$(c_1r_0 + q_0)C_0(1 + 2\bar{q}) < q < 1 \qquad \bar{q} < q \tag{2.34}$$

Proof: The proof is exactly the same as in Lemma 2.1, Chapter 7. However, relations (2.21) and (2.22) impose additional restriction on Q (see Lemma 1.6, Chapter 2) where

$$\sigma = \nu \qquad 0 < \eta < \theta^{-m}$$

and

$$Q < K^{(\nu-\bar{\nu})/\xi} \quad \blacksquare \tag{2.35}$$

2.5 LEMMA For $x_n \in G$, let

$$\|x_n\|_{\infty,p} < Aq^{-\alpha n} = K_n \tag{2.36}$$

$$\|Px_n\|_{\infty,0} \le p_0q^n \tag{2.37}$$

where $p_0 = \|Px_0\|_{\infty,0}$ and $\alpha > 1$, A are chosen so as to satisfy relations (2.17) and (2.18), Chapter 7. Then there exists a number $Q = Q_n$ such that

$$MQ_n^{-\mu}K^{\nu} \le \bar{q}p_0q^n \tag{2.38}$$

and

$$\|x_n + h_n\|_{\infty,p} < AQ^{-\alpha(n+1)} = K_{n+1} \tag{2.39}$$

h_n is a solution of the equation (2.28) with $x = x_n$, i.e.,

$$Ah_n + F'(x_n)h_n + Px_n + y_n = 0 \tag{2.40}$$

which satisfies relation (2.29) with $\sigma = \nu$ and (2.30).

Proof: The proof of Lemma 2.2, Chapter 7, carries over. ∎

2.5a LEMMA There exist ν with $\bar{\nu} < \nu < 1$ and μ, which satisfies (2.21) and such that for $K > 1$ and $Q > 1$ one can find $0 < \eta < 1$

and $\theta > 1$, which satisfy (2.23) and (2.22) provided

$$0 < \bar{k} < 1 \tag{2.41}$$

$$\bar{\nu} = \frac{\bar{p} - \rho}{p - \rho} \tag{2.42}$$

with ρ as in the proof of Lemma 1.6, Chapter 2, where

$$0 < \rho < \bar{p} \tag{2.43}$$

Proof: For the proof see Lemma 1.6, Chapter 2. ∎

Notice that m_2 in (2.22) and (2.9) can be replaced by $\bar{p}$.

We shall now construct the following iterative method. With x_0 as above, assume that $x_0, x_1, \ldots, x_n$ are known and satisfy the relations (2.36) and (2.37) of Lemma 2.5. Then put

$$x_{n+1} = z_n = x_n + h_n \tag{2.44}$$

where z_n and h_n are solutions of the equation (2.27) with $x = x_n$, $y = y_n$, and (2.28), respectively; i.e., z_n is an approximate solution of the linearized equation (2.1). It results from (2.39) that relation (2.36) holds true for $n + 1$, and relation (2.32) with $x = x_n$, $h = h_n$ implies that condition (2.37) is satisfied for $n + 1$.

2.6 LEMMA The following estimate holds.

$$\sum_{n=0}^{\infty} \|h_n\|_{\infty,s} \le \frac{N}{1 - q^{\beta}} \tag{2.45}$$

where $N = C[C_0(1 + \bar{q})p_0]^{1-\bar{s}}[A(q^{-\alpha} - 2)]^{\bar{s}}$, $\beta = [1 - \bar{s}(1 + \alpha)] > 0$, with $\bar{q}$ and q as in (2.34) and $\bar{s} = s/p$ such that

$$[\mu(1 - \nu) - \nu]^{-1} < \frac{1 - \bar{s}}{\bar{s}} = \frac{p - s}{s} \tag{2.46}$$

Proof: The proof is exactly the same as in Lemma 2.3, Chapter 7. ∎

Note that $\beta > 0$ implies $\alpha < (p - s)/s$ so that

$$[\mu(1 - \nu) - \nu]^{-1} < \alpha < \frac{p - s}{s} \tag{2.47}$$

2.1 THEOREM In addition to the assumptions (A_0) to (A_4), suppose that α satisfies (2.47) and A satisfies (1.19), Chapter 7. If

$$N(1 - q^{\beta}) < r_0$$

then equation (2.1) has a solution x such that

$$\|x_n - x\|_{\infty,s} \to 0 \qquad \text{as } n \to \infty$$

and $\{x_n\} \in G$.

Proof: The proof of Theorem 2.1, Chapter 7, carries over. ∎

For applications of the theory of C_0-semigroups and for the uniqueness of the solutions, see Section 4, Chapter 7.

3. A GENERAL EXISTENCE THEOREM FOR NONLINEAR EVOLUTION EQUATIONS VIA GLIM-III IN A SCALE OF BANACH SPACES

An iterative method is presented for solving general nonlinear evolution equations in nonreflexive Banach spaces. The method utilizes and further develops the ideas employed by Altman (1984c) and combines the concept of smoothing operators with the notion of elliptic regularization. As in Altman (1984c), the crucial idea which makes the method work is the concept of "convex approximate linearization" which is similar to the concept of "convex linearization" (see Altman, 1981). Moser's (1966) fundamental idea is still at work. However, it should be emphasized that Moser's method is designed for nonlinear operator equations and is not applicable to nonlinear evolution equations.

Let $\{X_j\}$ with $0 \le j \le p$ be a scale of Banach spaces with increasing norms such that $i < j$ implies $X_j \subset X_i$ and $\|x\|_j \ge \|x\|_i$, and let $0 < m_1 < m_2 < \bar{p} < p$.

(A_0) We assume that there exists a family of linear (smoothing) operators S_θ, $\theta \ge 1$, which satisfy the same conditions as in Section 1. Given $0 < b$, denote by $C(0,b;X_j)$ the Banach space of all continuous functions $x = x(t)$ defined on the interval $(0,b)$ with values in X_j and norm $\|x\|_{\infty,j} = \sup_t[\|x(t)\|_j: 0 \le t \le b]$, for $j = 0, m_1, m_2, \bar{p}, p$ and s to be determined later. Denote by $C^1(0,b;X_0)$ the vector space of all continuously differentiable functions from $(0,b)$ to X_0. Let $W_0 \subset X_s$ be an open ball with center $x_0 \in X_p$ and radius $r > 0$. Put $V_0 = W_0 \cap X_p$ and let V_s be the $\|\cdot\|_s$-closure of V_0.

Let G be a set of functions

$$x \in C(0,b;V_0(\|\cdot\|_p)) \cap C^1(0,b;X_0) \qquad \|x - x_0\|_{\infty,s} < r$$

with $x(0) = x_0$ and $\|x\|_{\infty,p} < \infty$.

Let $F\colon (0,b) \times V_s \to X_0$ be a nonlinear mapping and consider the Cauchy problem

$$Px(t) \equiv \frac{dx}{dt} + F(t,x) = 0 \qquad 0 \le t \le b \qquad x(0) = x_0 \tag{3.1}$$

We make the following assumptions.

(A_1) F is continuous and differentiable in the following sense.

$$\|x_n - x\|_{\infty,s} \to 0 \quad \text{implies} \quad \|F(\cdot,x_n) - F(\cdot,x)\|_{\infty,0} \to 0 \tag{3.2}$$

as $n \to \infty$, where $\{x_n\} \subset G$. For each $(t,x) \in (0,b) \times G$, there exists a linear operator $F'(t,x)$ such that

$$\varepsilon^{-1}\|F(\cdot, x + \varepsilon h) - F(\cdot,x) - \varepsilon F'(\cdot,x)h\|_{\infty,0} \to 0 \tag{3.3}$$

as $\varepsilon \to 0+$, where $h \in C(0,b;X_p)) \cap C^1(0,b;X_0)$.

There exists a constant $C > 0$ such that

$$\|F'(t,x)h\|_{m_1} \le C\|h\|_{m_2} \tag{3.4}$$

for all $(t,x) \in (0,b) \times V_0$ and $h \in X_{m_2}$, and

$$\|F'(t,x)y\|_0 \le C\|y\|_{m_1} \tag{3.5}$$

for all $(t,x) \in (0,b) \times V_0$ and $y \in X_{m_1}$.

(A_2) F' satisfies the relation

$$\|F(\cdot, x + h) - F(\cdot,x) - F1(\cdot,x)h\|_{\infty,0} \le M_0\|h\|_{\infty,0}^{2-\beta}\|h\|_{\infty,p}^{\beta} \qquad (3.6)$$

for some $M_0 > 0$, $0 \le \beta < 1$, and all $x \in G$.

(A_3) There exists a linear (regularizing) operator $L = L(\eta)$ such that

$$\|Lx\|_{m_1} \le C\|x\|_{m_2} \qquad (3.7)$$

for some constant $C > 0$.

The linearized equation

$$\frac{dz}{dt} + F'(t,x)z + \eta Lz + F(t,x) - F'(t,x)x = 0 \qquad (3.8)$$

$0 \le t \le b$, $z(0) = 0$, with small $\eta > 0$, has a solution $\bar{z}$ such that

$$\|\bar{z}\|_{\infty,i} \le C\eta^{-k}\|F(\cdot,x) - F'(\cdot,x)x\|_{\infty,m_2} \qquad i = m_2 \text{ or } \bar{p} \qquad (3.9)$$

for some $C > 0$ and $k > 0$ to be determined.

(A_4) There exists $C > 0$ such that $x \in G$ implies

$$\|F(\cdot,x) - F'(\cdot,x)x\|_{\infty,m_2} \le C\|x\|_{\infty,\bar{p}} \qquad (3.10)$$

(A_5) For $x \in G$ and g with $\|g\|_{\infty,0} < \infty$, if h is a solution of the equation

$$\frac{dh}{dt} + F'(t,x)h + g = 0 \qquad 0 \le t \le b \qquad h(0) = 0 \qquad (3.11)$$

then

$$\|h\|_{\infty,0} \le C\|g\|_{\infty,0} \qquad (3.12)$$

for some constant $C > 0$.

For $x \in G$, let $\bar{z}$ be a solution of equation (3.8) satisfying (3.9). Then we have the estimates:

3.1 LEMMA The following relations hold for $x \in G$ with $\|x\|_{\infty,p} < K$, $K > 1$, $0 < \eta < 1$.

$$\|(I - S_\theta)\frac{d\bar{z}}{dt}\|_{\infty,0} \le M_1\theta^{-m_1}\eta^{-k}K^{\bar{p}/p} \tag{3.13}$$

$$\|F'(\cdot,x)(I - S_\theta)\bar{z}\|_{\infty,0} \le M_2\theta^{-(m_2-m_1)}\eta^{-k}K^{\bar{p}/p} \tag{3.14}$$

$$\|z\|_{\infty,p} = \|S_\theta\bar{z}\|_{\infty,p} \le M_3\theta^{p-\bar{p}}\eta^{-k}K^{\bar{p}/p} \tag{3.15}$$

for some M_1, M_2, M_3; $z = S_\theta\bar{z}$.

Proof: We get from (3.8) and (3.9)

$$\begin{aligned}\|(I - S_\theta)\frac{d\bar{z}}{dt}\|_0 &= \|(I - S_\theta)[F'(t,x)\bar{z} + \eta L\bar{z} + F(t,x) - F'(t,x)x]\|_0 \\ &\le C\theta^{-m_1}(\|F'(t,x)\bar{z}(t)\|_{m_1} + C\|\bar{z}\|_{m_2} \\ &\quad + \|F(t,x) - F'(t,x)x\|_{m_1}) \\ &\le C\theta^{-m_1}(C\|\bar{z}\|_{m_2} + C\|\bar{z}\|_{m_2} + C\|x\|_{\infty,\bar{p}}) \\ &\le C\theta^{-m_1}3C\eta^{-k}\|x\|_{\infty,\bar{p}} \le M_1\theta^{-m_1}\eta^{-k}K^{\bar{p}/p}\end{aligned}$$

since

$$\|x\|_{\infty,\bar{p}} \le C\|x\|_{\infty,0}^{1-\bar{p}/p}\|x\|_{\infty,p}^{\bar{p}/p} \le Cr^{1-\bar{p}/p}K^{\bar{p}/p}$$

and

$$\|\bar{z}\|_{\infty,\bar{p}} \le C_1\eta^{-k}\|x\|_{\infty,\bar{p}} \tag{3.16}$$

We have

$$\begin{aligned}\|F'(t,x)(I - S_\theta)\bar{z}(t)\|_0 &\le C\|(I - S_\theta)\bar{z}(t)\|_{m_1} \\ &\le C\theta^{-(m_2-m_1)}\|\bar{z}\|_{\infty,m_2}\end{aligned}$$

and (3.14) follows from (3.16).

By virtue of (A_0), we get

$$\|S_\theta\bar{z}(t)\|_p \le C\theta^{p-\bar{p}}\|z\|_{\infty,\bar{p}}$$

and (3.15) results from (3.16). ■

Consider the linearized equation

$$\frac{dz}{dt} + F'(t,x)z + F(t,x) - F'(t,x)x = 0 \qquad (3.17)$$
$$0 \le t \le b, \ z(0) = x_0$$

and put $z = S_\theta \bar{z}$, where $\bar{z}$ is a solution of the linearized equation (3.8) satisfying relation (3.9).

3.2 LEMMA The following estimate holds.

$$\left\|\frac{dz}{dt} + F'(t,x)z + F(\cdot,x) - F'(\cdot,x)x\right\|_{\infty,0}$$
$$\le M_4(\theta^{-m} + \eta)\eta^{-k}K^{\bar{p}/p} \qquad (3.18)$$

for $x \in G$ with $\|x\|_{\infty,p} < K$, some M_4, where

$$m = \min(m_1, m_2 - m_1) \qquad (3.19)$$

Proof: We have

$$\left\|\frac{dz}{dt} + F'(t,x)z + F(\cdot,x) - F'(\cdot,x)x\right\|_{\infty,0}$$
$$\le \left\|\frac{d\bar{z}}{dt} + F'(t,x)\bar{z} + \eta L\bar{z} + F(\cdot,x) - F'(\cdot,x)x\right\|_{\infty,0}$$
$$+ \left\|(I - S_\theta)\frac{d\bar{z}}{dt}\right\|_{\infty,0} + \|F'(\cdot,x)(I - S_\theta)\bar{z}\|_{\infty,0} + \eta\|L\bar{z}\|_{\infty,0}$$

and (3.18) follows, by virtue of Lemma 3.1, (3.8), (3.9), and (3.19). ■

Now put $i = \bar{p}$ in (3.9)

$$Q = \theta^{p-\bar{p}} \qquad (3.20)$$

$$\mu = \frac{m}{p - \bar{p}}$$

Let $0 < \alpha < 1$, and choose $0 < \eta < 1$ so as to satisfy the relations

$$\eta^{-k}K^{\bar{p}/p} < K^{1+\alpha} \quad \text{and} \quad \eta < Q^{-\mu} \qquad (3.22)$$

Then relation (3.18) can be written as

$$\left\|\frac{dz}{dt} + F'(t,x)z + F(\cdot,x) - F'(\cdot,x)x\right\|_{\infty,0} \le 2M_4 Q^{-\mu} K^{1+\alpha} \tag{3.23}$$

since $\theta^{-m} = Q^{-\mu}$ by (3.20), (3.21), and (3.22).

3.3 LEMMA The linearized equation (3.17) admits approximate solutions of order (μ,ν,σ) with $\nu = \sigma = 1 + \alpha$ in the sense of Definition 3.1, Chapter 7.

Proof: The proof follows from (3.23), (3.15), and (3.22). ■

Thus we have shown that if $\bar{z}$ is a solution of the equation (3.8) satisfying (3.9), then $z = S_\theta \bar{z}$ is an approximate solution of order (μ,ν,σ) of the linearized equation (3.17); that is, there exist a function y and a constant M such that

$$\frac{dz}{dt} + F'(t,x)z + F(\cdot,x) - F'(\cdot,x)x + y = 0 \tag{3.24}$$

$0 \le t \le b$, $z(0) = x_0$, with

$$\|y\|_{\infty,0} \le MQ^{-\mu}K^{1+\alpha} \tag{3.25}$$

and

$$\|z\|_{\infty,p} = \|S_\theta \bar{z}\|_{\infty,p} \le MQK^{1+\alpha} \tag{3.26}$$

provided that $x \in G$ with $\|x\|_{\infty,p} < K$, $K > 1$, and η in (3.8) is chosen so as to satisfy (3.22).

Now put $z = x + h$. Then it follows from (3.24) that h is a solution of the equation

$$\frac{dh}{dt} + F'(t,x)h + Px + y = 0 \qquad 0 \le t \le b \qquad h(0) = 0 \tag{3.27}$$

In order to construct the iterative method in question, let $K_{n+1} = K_n^{\tau+\alpha}$, where $1 < \tau < 2$, $0 < \alpha < 1$, K_0 being sufficiently large large, and put

$$x_0(t) \equiv x_0 = 0 \qquad x_{n+1} = z_n = x_n + h_n \qquad x_n \in G \tag{3.28}$$

where z_n is a solution of (3.24), and Q_n from (3.25) and (3.26) is chosen so as to satisfy the induction assumptions

$$\|x_n\|_{\infty,p} < K_n \qquad x_n \in G \tag{3.29}$$

$$\|Px_n\|_{\infty,0} \le K_n^{-\lambda} \tag{3.30}$$

for some $\lambda > 0$ to be determined. To verify the induction assumption (3.29) we use relations (3.28) and (3.26) with $z = z_n$, and choose Q so as to satisfy

$$2MQK_n^{1+\alpha} < K_{n+1} = K_n^{\tau+\alpha} \tag{3.31}$$

Then we get

$$\|x_{n+1}\|_{\infty,p} = \|z_n\|_{\infty,p} < MQK_n^{1+\alpha} < K_{n+1} \tag{3.32}$$

To verify the induction assumption (3.30) we use relation (3.6) with $x = x_n$, $h = h_n$, where h_n satisfies (3.27) with $y = y_n$, and the following identity:

$$\begin{aligned} Px_{n+1} &= P(x_n + h_n) \\ &= [F(\cdot,\ x_n + h_n) - F(\cdot, x_n) - F'(\cdot, x_n)h_n] \\ &\quad + [dh_n/dt + F'(t, x_n)h_n + Px_n + y_n] - y_n \end{aligned}$$

Hence,

$$\|Px_{n+1}\|_{\infty,0} \le M\|h_n\|_{\infty,0}^{2-\beta}\|h_n\|_{\infty,p}^{\beta} + \|y_n\|_{\infty,0} \tag{3.33}$$

But (3.12) with $g = Px_n + y_n$ implies

$$\begin{aligned} \|h_n\|_{\infty,0} &\le C(\|Px_n\|_{\infty,0} + \|y_n\|_{\infty,0}) \\ &\le C(K_n^{-\lambda} + MQ^{-\mu}K_n^{1+\alpha}) \end{aligned}$$

by (3.30) and (3.25). Hence,

$$\|h_n\|_{\infty,0} \le 2CK_n^{-\lambda} \tag{3.34}$$

provided Q is chosen so as to satisfy

$$MQ^{-\mu}K_n^{1+\alpha} < K_n^{-\lambda} \tag{3.35}$$

Since $z_n = x_n + h_n$, we get

$$\|h_n\|_{\infty,p} \le \|x_n\|_{\infty,p} + \|z_n\|_{\infty,p}$$
$$\le K_n + MQK_n^{1+\alpha}$$

Hence,

$$\|h_n\|_{\infty,p} < 2MQK_n^{1+\alpha} < K_{n+1} \tag{3.36}$$

by (3.31). Relations (3.33), (3.34), and (3.36) yield

$$\|Px_{n+1}\|_{\infty,0} \le M(2C)^{2-\beta}K_n^{-\lambda(2-\beta)}(2M)^{\beta}(QK_n^{1+\alpha})^{\beta} + MQ^{-\mu}K_n^{1+\alpha}$$
$$\le c[K_n^{-\lambda(2-\beta)}(QK_n^{1+\alpha})^{\beta} + Q^{-\mu}K_n^{1+\alpha}]$$

where $c > M$ is some constant. Thus the induction assumption (3.30) can be verified if Q can be chosen so as to satisfy

$$c[K_n^{-\lambda(2-\beta)}(QK_n^{1+\alpha})^{\beta} + Q^{-\mu}K_n^{1+\alpha}] < K_{n+1}^{-\lambda} \tag{3.37}$$

Therefore, by virtue of (3.31), (3.35), (3.36), and (3.37), it suffices to show that the following inequalities can be satisfied for some $Q > 1$:

$$cK_n^{1+\alpha}Q < K_{n+1} \tag{3.38}$$

$$c(K_n^{1+\alpha}Q)^{\beta}K_n^{-\lambda(2-\beta)} < K_{n+1}^{-\lambda} \tag{3.39}$$

$$cK_n^{1+\alpha}Q^{-\mu} < K_{n+1}^{-\lambda} \tag{3.40}$$

and, in addition, relations (3.22) with $\varepsilon = \varepsilon_n$ and $K = K_n$ imply

$$Q < K_n^{(1+\alpha-\bar{p}/p)/\mu k} \tag{3.41}$$

(A_6) Let $0 < \alpha < \alpha_0 < 1$, $\lambda < \mu$; $\mu = m/(p - m_2) > 1$, which implies $p < \min(2m_2 - m_1, m_2 + m_1)$, be such that

$$0 < 2\lambda < [\mu(1 - \alpha_0) - (1 + \alpha_0)]/(1 + \alpha_0) \tag{3.42}$$

$$\mu > (1 + \alpha_0)/(1 - \alpha_0) > 1 \tag{3.43}$$

and let β,τ be such that

$$0 < \beta < \mu\lambda(\alpha_0 - \alpha)[(1 + \alpha)(1 + \mu) + \lambda(2 + \mu)]^{-1} < 1 \tag{3.44}$$

$$1 < (\mu - \lambda)^{-1}[1 + \alpha(1 + \lambda) + \mu] < \tau < 2 - \alpha_0 < 2 - \alpha \tag{3.45}$$

where one can put, e.g., $\alpha_0 = (\mu - 1)/(\mu + 3)$ in (3.42) to (3.45).

It is shown in Section 3, Chapter 1, that relations (3.42) to (3.44) imply the existence of τ satisfying (3.45), and Q can be found to satisfy relations (3.38) to (3.40) if K_0 is sufficiently large. But (3.38) implies that $Q < K_n^{\tau-1}$. Hence, relation (3.41) holds if k is such that

$$\tau - 1 < \frac{(1 + \alpha - \bar{p}/p)}{\mu k} \tag{3.46}$$

or if

$$k < \frac{(1 + \alpha - \bar{p}/p)}{(\tau - 1)\mu} \tag{3.47}$$

Then $k \geq 1$ if $1 + \alpha - \bar{p}/p > (\tau - 1)\mu$ or if $\bar{p}/p < 1 + \alpha - (\tau - 1)\mu$, which implies $1 + \alpha > (\tau - 1)\mu$ or $\mu < (1 + \alpha)/(\tau - 1)$. Hence, by (3.43), we get $(1 + \alpha_0)/(1 - \alpha_0) < (1 + \alpha)/(\tau - 1)$, which implies $\tau < (1 + \alpha)(1 - \alpha_0)/(1 + \alpha_0) + 1$. Hence, by (3.45), we get the relation $(1 + \alpha(1 + \lambda) + \mu)/(\mu - \lambda) < (1 + \alpha)(1 - \alpha_0)/(1 + \alpha_0) + 1$, and consequently, we obtain the relation $(1 + \lambda)(1 + \alpha_0)/(1 - \alpha_0) < \mu - \lambda$ or $2\lambda < \mu(1 - \alpha_0) - (1 + \alpha_0) = (1 + \alpha_0)[\mu(1 - \alpha_0)/(1 + \alpha_0) - 1]$ which is satisfied by virtue of (3.42).

At this point let us observe that if $\alpha = 0$, which is the case studied in Altman (1984a), then relation (3.47) yields

$$k < \frac{(1 - \bar{p}/p)}{(\tau - 1)\mu} \tag{3.48}$$

which implies $k < 1$. In fact, if $k \geq 1$, then $\bar{p}/p < 1 - (\tau - 1)\mu$. Hence, $1 - (\tau - 1)\mu > 0$ or $\tau < 1 + 1/\mu$. On the other hand [see (3.36), Chapter 1], we have $(1 - (\lambda + 1)/(\mu + 1))^{-1} < \tau$ which leads to a contradiction. Therefore, the introduction of $0 < \alpha < 1$ has certain advantage over the case of $\alpha = 0$, enhancing the applicability of the method in question.

We are now in a position to prove the following existence theorem.

3.1 THEOREM Suppose that assumptions (A_0) to (A_6) are satisfied. Then there exists a constant $K_0(M,\beta,\mu,\lambda) > 1$ such that

$$\|F(\cdot,0)\|_{\infty,0} < K_0^{-\lambda}$$

$$C(2C)^{1-s/p} \sum_{n=0}^{\infty} K_n^{-\delta} < r \tag{3.49}$$

where

$$\frac{s}{p} < \frac{\lambda}{\lambda + 2} \tag{3.50}$$

yielding $\delta = (1 - s/p)\lambda - \tau s/p > 0$, and $K_{n+1} = K^{\tau+\alpha}$ with τ as in (3.45), imply that the equation (3.1) has a solution x which is a limit of $\{x_n\}$ determined by (3.28) and

$$\|x_n - x\|_{\infty,s} \to 0 \qquad \text{as } n \to \infty \tag{3.51}$$

Proof: We get from (A_0), (3.34), and (3.36),

$$\begin{aligned}\|h_n\|_{\infty,s} &\le C\|h_n\|_{\infty,0}^{1-s/p}\|h_n\|_{\infty,p}^{s/p} \\ &\le C(2CK_n^{-\lambda})^{1-s/p}K_n^{\tau s/p} = C(2C)^{1-s/p}K_n^{-\delta}\end{aligned}$$

with δ as in (3.49). Hence, it follows from (3.49) that $\|x_n\|_{\infty,s} < r$, and (3.51) holds for x being the limit of $\{x_n\}$. It results from (3.2) and (3.30) that x is a solution of equation (3.1). It is clear that relation (3.49) is satisfied if K_0 is sufficiently large. This completes the proof. ∎

3.1 REMARK Lemma 3.4 and Remarks 3.3 and 3.4, Chapter 2, also apply to this section. Then $i = m_2$ in (3.9).

4. SOLVING NONLINEAR EVOLUTION EQUATIONS VIA LINEARIZED EVOLUTION EQUATIONS

The vast majority of nonlinear problems can be reduced to solving the appropriate linearized equations. This is a rather natural approach, and Kato (1975) used this method for solving quasilinear evolution equations. However, in case of general nonlinear evolution equations in Banach spaces, it was not clear how to do it. Fortunately, our theory, which is in general independent of the theory of C_0-semigroups, also presents a method of solving nonlinear evolution equations via linearized evolution equations. This fact is quite important from the point of view of applications to nonlinear partial differential equations. In this way one can make use of the theory of C_0-semigroups, which is more powerful and advanced than the theory of nonlinear semigroups, which is rather limited in scope. Nevertheless, many interesting results have been obtained for nonlinear evolution equations by the latter method. For references, see Barbu (1976).

Denote by $G(X_0)$ the set of all negative infinitesimal generators of C_0-semigroups $\{U(t)\}$ in X_0. If $-A \in G(X_0)$, then $\{U(t)\} = \{e^{-tA}\}$, $0 \le t < \infty$, is the semigroup generated by $-A$. For $x \in G$, put $A(t,x(t)) = F'(t,x(t))$. We assume that A is a function from $[0,b] \times V_0$ into $G(X_0)$, and $A(t) = A(t,x(t))$ is stable in X_0 (see Kato, 1970; 1973; 1975; Yosida, 1968; Tanabe, 1979). We also assume that X_{m_2} being dense in X_0 is admissible and preserves the stability of $A(t)$, and the evolution operators $U(t,s;x)$ generated by $A(t,x(t))$ exist and

$$\|U(t,s;x)\|_{X_0} \le M_0 \quad \text{and} \quad \|U(t,s;x)\|_{X_{m_2}} \le M_2 \tag{4.1}$$

for all $x \in G$ and some $M_0, M_2 > 0$. These assumptions imply that (1.11) and (1.13) hold with $L = 0$ and $\bar{k} \ge 0$ arbitrary. The same holds for Sections 2 and 3.

EXAMPLE 1 Consider the nonlinear equation

$$u_t + F(t,u,u_x) = 0 \qquad 0 \le t \le b \qquad -\infty < x < \infty \tag{4.2}$$

For $x \in G$ with appropriate choice of s, it follows from Kato's (1975) argument that the linearized equations for (4.2) satisfy (4.1) with $X_{m_2} = H^r(-\infty,\infty)$, $r \geq 2$, since $\|u\|_{\infty,s} < r_0$.

EXAMPLE 2 Korteweg-de Vries nonlinear equation

$$u_t + u_{xxx} + F(t,u,u_x) = 0 \qquad 0 \leq t \leq b \qquad -\infty < x < \infty \tag{4.3}$$

It follows from Kato's (1976) argument and Remark 2.1 that the linearized equations for (4.3) satisfy (4.1) if $x_{m_2} = H^4(-\infty,\infty)$ with $r \geq 3$, where $u \in G$ and s is properly chosen.

EXAMPLE 3 Consider the nonlinear system

$$\frac{du}{dt} + F(t,x,u,u_x) = 0 \qquad 0 \leq t \leq b$$

where $u = u(t,x) = u(t,x), \ldots, u_N(t,x))$, $x \in R^m$, $F = (F_1,\ldots,F_N)$. Suppose that the linearized systems for (4.3) satisfy Kato's (1975) assumptions for symmetric hyperbolic systems. Then (4.1) is satisfied with x replaced by u provided that s is properly chosen.

4.1 REMARK The evolution operators $U(t,s;x)$ in (4.1) can be replaced by their approximations $U_m(t,s;x)$ generated by the appropriate step functions.

9

Elliptic Regularization Without Smoothing Operators for Nonlinear Evolution Equations

1. ELLIPTIC REGULARIZATION AND GLIM-I FOR NONLINEAR EVOLUTION EQUATIONS

Approximate solutions of the linearized equations employed in Section 1, Chapter 7, can be constructed without making use of smoothing operators. But then stronger conditions have to be imposed on the regularizing operator L.

Let $Z \subset H \subset E \subset Y \subset X$ be Banach spaces with norms $\|\cdot\|_X \le \|\cdot\|_Y \le \|\cdot\|_E \le \|\cdot\|_H \le \|\cdot\|_Z$ and let $W_0 \subset Y$ be an open ball with center $x_0 \in Z$ and radius $r_0 > 0$. Put $V_0 = W_0 \cap Z$ and let V_1 be the $\|\cdot\|_Y$-closure of V_0. As in Section 1, Chapter 7, the nonlinear evolution equation is

$$Px(t) \equiv \frac{dx}{dt} + F(t,x) + f(t,x) = 0 \tag{1.1}$$

$0 \le t \le b$, $x(0) = x_0$, where $F,f\colon\ (0,b) \times V_1 \to X$. If the domain of P is D, then we put $V_0 = D \cap W_0 \cap Z$.

We make the following assumptions.

(A_0) In addition to (1.0), Chapter 7, the following condition holds for some $C > 0$, $0 < \nu < 1$,

$$\|x\|_H \le C\|x\|_X^{1-\nu}\|x\|_Z^{\nu}$$

(A_1) As in Chapter 7, we assume F,f are continuous. F is differentiable and F',f satisfy assumptions (A_1) of Section 1, Chapter 7.

(A_2) There exists a linear (regularizing) operator $L = L(\eta)$ such that

$$\|Lz\|_X \leq C_1\|z\|_Z \tag{1.2}$$

for some $C_1 > 0$. The linearized equation

$$\frac{dz}{dt} + F'(t,x)z + \eta Lz + F(t,x) - F'(t,x)x + f(t,x) = 0 \tag{1.3}$$

$0 \leq t \leq b$, $z(0) = x_0$, and $0 < \eta < 1$ has a solution z such that

$$\|z(t)\|_Z \leq C\eta^{-1/\omega}\|F(t,x) - F'(t,x)x + f(t,x)\|_E \tag{1.4}$$

for some $C > 0$, $\omega > 1$, and all $x = x(t) \in G$.

(A_3) There exists a constant $C > 0$, such that for all $x = x(t) \in G$,

$$\|F(t,x) - F'(t,x)x + f(t,x)\|_E \leq C\|x\|_H \tag{1.5}$$

(A_4) For $x \in G$, if h is a solution of the equation

$$\frac{dh}{dt} + F'(t,x)h + g = 0 \tag{1.6}$$

then

$$\|h\|_{\infty,X} \leq C_0\|g\|_{\infty,X} \tag{1.7}$$

for some constant $C_0 > 0$.

1.1 LEMMA The linearized equation

$$\frac{dz}{dt} + F'(t,x)z + F(t,x) - F'(t,x)x + f(t,x) = 0 \qquad 0 \leq t \leq b \quad z(0) = x_0 \tag{1.8}$$

admits approximate solutions of order (μ,ν,σ) in the sense of Definition 1.1, Chapter 7, with $\sigma = \nu$, $\mu = \omega - 1$.

Proof: For $x \in G$ with $\|x\|_{\infty,Z} < K$, $K > 1$, and $\eta = Q^{-\omega}$ with $Q > 1$, let z be a solution of the equation (1.3). Then it follows from relations (1.4), (1.5), and (A_0) that

$$\|z\|_{\infty,Z} \leq MQK^{\nu} \tag{1.9}$$

for some constant $M > 0$, $0 < \nu < 1$. On the other hand, we get from (1.3):

$$\left\|\frac{dz}{dt} + F'(t,x)z + F(t,x) - F'(t,x)x + f(t,x)\right\|_{\infty,X}$$

$$\leq \eta\|Lz\|_{\infty,X} \leq C_1Q^{-\omega}\|z\|_{\infty,Z} \leq M_1Q^{-(\omega-1)}K^{\nu}$$

Thus $K > 1$, $Q > 1$, and $x \in G$ with $\|x\|_{\infty,Z} < K$ imply the existence of a function z satisfying (1.9) and a function y with

$$\|y\|_{\infty,X} \leq MQ^{-\mu}K^{\nu} \tag{1.10}$$

and such that

$$\frac{dz}{dt} + F'(t,x)z + F(t,x) - F'(t,x) + f(t,x) + y = 0 \qquad 0 \leq t \leq b \qquad z(0) = x_0 \tag{1.11}$$

Now put $z = x + h$. Then we get from (1.11):

$$\frac{dh}{dt} + F'(t,x)h + Px + y = 0 \qquad 0 \leq t \leq b \qquad h(0) = 0 \tag{1.12}$$

Now in the same way as in Chapter 7, we can construct the following method of contractor directions. Put $x_0(t) \equiv x_0$,

$$x_{n+1} = x_n + \varepsilon h_n \qquad t_{n+1} = t_n + \varepsilon_n \qquad t_0 = 0 \tag{1.13}$$

where h_n is a solution of the equation (1.12) with $x = x_n$, $y = y_n$, where $x_n \in G$. The induction assumptions for $\{x_n\}$ are the same as in Section 1, Chapter 7, i.e., (1.27) and (1.28), and $\alpha > 1$, $A > 0$ satisfy (1.17) to (1.19), and (1.32). The choice of $\{\varepsilon_n\}$ is exactly the same as in Section 1, Chapter 7, and the estimate (1.31) is valid. ∎

1.1 THEOREM In addition to the assumptions (A_0) to (A_4) suppose that conditions (1.12) and (1.17) to (1.19), Chapter 7, hold, and b' is such that condition (1.31), Chapter 7, is satisfied. Then the equation (1.1) with b replaced by b' has a solution x satisfying (1.33) and (1.34), Chapter 7, where $\{x_n\}$ is determined by (1.13).

Proof: The proof of Theorem 1.1, Chapter 7, carries over. ■

2. ELLIPTIC REGULARIZATION AND GLIM-II FOR NONLINEAR EVOLUTION EQUATIONS

The discussion in Section 1 concerning elliptic regularization without smoothing operators also applies to GLIM-II. In this case F is to satisfy assumption (A_1) of Section 2, Chapter 7, and f is subject to the conditions mentioned there in Remark 2.1. The remaining hypotheses are the same as in Section 1 of this chapter. The method is defined by the formula (1.3), where $\varepsilon_n = 1$ and $t_n = 0$ for all n. Based on the proof of Theorem 2.1, Chapter 7, an existence and convergence theorem can be proved for GLIM-II.

3. ELLIPTIC REGULARIZATION AND GLIM-III FOR NONLINEAR EVOLUTION EQUATIONS

By imposing stronger conditions on the regularizing linear operator L, an existence theorem can be proved without making use of smoothing operators. Let $Z \subset E \subset Y \subset X$ be Banach spaces with norms $\|\cdot\|_X \le \|\cdot\|_Y \le \|\cdot\|_E \le \|\cdot\|_Z$ and let $W_0 \subset Y$ be an open ball with center $x_0 \in Z$ and radius $r > 0$. Put $V_0 = W_0 \cap Z$ and let V_1 be the $\|\cdot\|_Y$-closure of V_0. Using the same notation as in Section 1, let G be a set of functions, $x \in C(0,b;V_0(\|\cdot\|_Z)) \cap C^1(0,b;X)$ with $x(0) = x_0$, $\|x - x_0\|_{\infty,Y} < r$, and $\|x\|_{\infty,Z} < \infty$.

Let $F\colon\ (0,b) \times V_1 \to X$ be a nonlinear mapping and consider the Cauchy problem

$$Px(t) \equiv \frac{dx}{dt} + F(t,x) = 0 \qquad 0 \le t \le b \qquad x(0) = x_0 \tag{3.1}$$

We make the following assumptions.

(A_0) There exist a number $0 < \bar{s} < 1$ and a constant $C > 0$ such that

$$\|x\|_Y \le C\|x\|_X^{1-\bar{s}}\|x\|_Z^{\bar{s}}$$

(A_1) F is continuous and differentiable in the following sense.

$$\|x_n = x\|_{\infty,Y} \to 0 \qquad \text{implies} \qquad \|F(\cdot,x_n) - F(\cdot,x)\|_{\infty,X} \to 0 \tag{3.2}$$

as $n \to \infty$, where $\{x_n\} \subset G$. For each $(t,x) \in (0,b) \times G$, there exists a linear operator $F'(t,x)$ such that

$$\varepsilon^{-1}\|F(\cdot,\ x + \varepsilon h) - F(\cdot,x) - \varepsilon F'(\cdot,x)h\|_{\infty,X} \to 0 \tag{3.3}$$

as $\varepsilon \to 0+$, where $h \in C(0,b);Z) \cap C^1(0,b;X)$.

(A_2) F' satisfies the relation

$$\|F(\cdot,\ x + h) - F(\cdot,x) - F'(\cdot,x)h\|_{\infty,X} \le M_0\|h\|_{\infty,X}^{2-\beta}\|h\|_{\infty,Z}^{\beta} \tag{3.4}$$

for some M_0, $0 \le \beta < 1$, and all $x \in G$.

(A_3) There exists a linear (regularizing) operator $L = L(\eta)$ such that

$$\|Lx\|_X \le C_1\|x\|_Z \tag{3.5}$$

for some constant $C_1 > 0$. The linearized equation

$$\frac{dz}{dt} + F'(t,x)z + \eta Lz + F(t,x) - F'(t,x)x = 0 \tag{3.6}$$

$0 \le t \le b$, $z(0) = x_0$, with small $\eta > 0$ has a solution z such that

$$\|z\|_{\infty,Z} \le C\eta^{-1/\bar{\mu}}\|F(\cdot,x) - F'(\cdot,x)x\|_{\infty,E} \tag{3.7}$$

for some $C > 0$, $\bar{\mu} > 1$, and all $x \in G$.

(A_4) There exists a constant $C > 0$ such that $x \in G$ implies

$$\|F(\cdot,x) - F'(\cdot,x)x\|_{\infty,E} \le C\|x\|_{\infty,Z} \tag{3.8}$$

(A_5) For $x \in G$ and g with $\|g\|_{\infty,X} < \infty$, if h is a solution of the equation

$$\frac{dh}{dt} + F'(t,x)h + g = 0 \qquad 0 \le t \le b \qquad h(0) = 0 \tag{3.9}$$

then

$$\|h\|_{\infty,X} \le C\|g\|_{\infty,X} \tag{3.10}$$

for some constant $C > 0$.

(A_6) We also assume that $\lambda > 0$ is such that

$$\eta + 1 < \frac{1}{2} (\mu + 1) \tag{3.11}$$

$$0 < \beta < \left(\frac{\lambda}{\lambda + 1}\right)\left(\frac{\mu}{\mu + 1}\right)\left(1 - 2\,\frac{\lambda + 1}{\mu + 1}\right) \tag{3.12}$$

with β as in (3.4).

3.1 LEMMA The linearized equation

$$\frac{dz}{dt} + F'(t,x)z + F(t,x) - F'(t,x)x = 0 \qquad 0 \le t \le b \qquad z(0) = 0 \tag{3.13}$$

admits approximate solutions of order (μ,ν,σ) with $\nu = \sigma = 1$ in the sense of Definition 3.1, Chapter 7.

Proof: Let z be a solution of equation (3.6) with $\|x\|_{\infty,p} < K$ and $K > 1$, where $\eta = Q^{-\bar{\mu}}$ for $Q > 1$. Then relations (3.7) and (3.8) imply

$$\|z\|_{\infty,Z} < MQK \tag{3.14}$$

where $M > 0$ is some constant. On the other hand, we get from (3.6),

$$\left\|\frac{dz}{dt} + F'(t,x)z + F(t,x) - F'(t,x)x\right\|_{\infty,X}$$

$$= \eta\|Lz\|_{\infty,X} \le C_1Q^{-\mu}\|x\|_{\infty,Z} < MQ^{-\mu}K \qquad \mu = \bar{\mu} - 1$$

by virtue of (3.5). Hence, for $x \in G$, we get

$$\left\|\frac{dz}{dt} + F'(t,x)z + F(t,x) - F'(t,x)x\right\|_{\infty,X} \le MQ^{-\mu}K \tag{3.15}$$

for some $M > 0$. Relation (3.15) yields the existence of a function y such that

$$\frac{dz}{dt} + F'(t,x)z + F(t,x) - F'(t,x)x + y = 0 \tag{3.16}$$

$0 \le t \le b$, $z(0) = x_0$, with

$$\|y\|_{\infty,X} < MQ^{-\mu}K \tag{3.17}$$

Now put $z = x + h$. Then we get from (3.16),

$$\frac{dh}{dt} + F'(t,x)h + Px + y = 0 \qquad 0 \le t \le b \qquad h(0) = 0 \tag{3.18}$$

Next one can construct the following iterative process,

$$x_0(t) \equiv x_0 \qquad x_{n+1} = z_n = x_n + h_n \tag{3.19}$$

where z_n, h_n are solutions of the equations (3.16) and (3.19), respectively. The induction assumptions are the same as in (3.29) and (3.30), Chapter 8, i.e.,

$$\|x_n\|_{\infty,Z} < K_n \qquad x_n \in G$$

$$\|Px_n\|_{\infty,X} \le K_n^{-\lambda}$$

with λ from (2.11), and $K_{n+1} = K_n^{\tau}$, where

$$1 < \left(1 - \frac{\lambda + 1}{\mu + 1}\right)^{-1} < \tau < 2 \qquad \blacksquare$$

3.1 THEOREM Suppose that assumptions (A_0) to (A_6) are satisfied. Then there exists a constant $K_0(M,\beta,\mu,\lambda) > 1$ such that

$$\|F(\cdot,x_0)\|_{\infty,X} < K_0^{-\lambda}$$

with λ from (3.11), $\|x_0\|_{\infty,Z} < K_0$,

$$C(2C)^{1-\bar{s}} \sum_{n=0}^{\infty} K_n^{-\delta} < r$$

where

$$\bar{s} < \frac{\lambda}{\lambda + 2}$$

and $\delta = (1 - \bar{s})\lambda - \tau\bar{s} > 0$ imply that equation (3.1) has a solution x which is a limit of $\{x_n\}$ determined by (3.19) and

$$\|x_n - x\|_{\infty,Y} \to 0 \qquad \text{as } n \to \infty$$

Proof: The proof is similar to that of Theorem 3.2, Chapter 7, except for some details. $\blacksquare$

3.2 THEOREM Suppose that assumptions A_0-A_6 hold, with (3.7) replaced by

$$\|z\|_{\infty,Z} \le C\varepsilon^{-1/\mu}K^{\alpha}\|F(\cdot,x) - F'(\cdot,x)x\|_{\infty,E} \tag{3.20}$$

where $\|x\|_{\infty,Z} < K$ and $0 < \alpha < 1$ is determined by (A_4) of Section 3, Chapter 7, and also (A_6) is replaced by the same (A_4) of Section 3, Chapter 7. Then the statement of Theorem 3.1 with $K_{n+1} = K_n^{\tau+\alpha}$ holds true.

Proof: The proof follows from Lemma 3.1 with $\nu = \sigma = 1 + \alpha$ and is similar to that of Theorem 3.1, Chapter 7, except for some details. ∎

It follows from Theorem 3.2 that by introducing a parameter $0 < \alpha < 1$, the method discussed above has a certain advantage. In fact, the estimate (3.7) can be replaced by the weaker one, i.e., (3.20).

Let us mention one basic difference between the method of Section 3, Chapter 8, which uses smoothing operators, and the one of this section, without smoothing operators. We have in the second method the exponent $\bar{k} = 1/\mu < 1$, whereas $\bar{k} \ge 1$ is admissible in Theorem 3.1, Chapter 8. Moreover, in both estimates (3.7) and (3.20), the norm on the left side must be stronger than the one on the right side, which is not the case for the estimate (3.9), Chapter 8.

10

Strongly Quasilinear Evolution Equations

Consider the quasilinear evolution equation

$$\frac{dx}{dt} + A(t,x)x - f(t,x) = 0 \qquad 0 \le t \le b \qquad x(0) = x_0 \tag{0.1}$$

where $A(t,x)h$ is linear in h, and assumes values in a Banach space X and so does dx/dt. Kato (1975) was the first to prove a general existence and uniqueness theorem for (0.1). He then successfully applied his outstanding result to partial differential equations in mathematical physics. His theory is based on C_0-semigroups and applies only to reflexive Banach spaces X. A generalization of the theory to nonreflexive Banach spaces X is given in Altman (1981). It is based on a different method [the method of contractor directions (Altman, 1980)] and is independent of C_0-semigroups, although a limited use of them is still a possibility. The following assumption is crucial to both theories.

(K) There exists an open ball, $W_0 \subset Z \subset X$, and a Banach space, Z, such that the linearized equations for (0.1) have solutions that are commonly bounded in Z for all $x(\cdot) \subset W_0$. However, in general, this assumption can be invalid. For example, consider a strictly nonlinear evolution equation

$$\frac{dx}{dt} + F(t,x) = 0 \qquad 0 \le t \le b \qquad x(0) = x_0 \tag{0.2}$$

and the corresponding quasilinear evolution equation

$$\frac{dx}{dt} + F'(t,x)x - f(t,x) = 0 \qquad 0 \le t \le b \qquad x(0) = x_0 \qquad (0.3)$$

where $F'(\cdot,x)$ is an abstract derivative of $F(\cdot,x)$ and $F'(t,x)h$ is linear in h. It is clear that for quasilinear equations (0.3), the crucial assumption (K) mentioned above may not be satisfied. This is the motivation for the introduction of a class of strongly quasilinear evolution equations. Thus strongly quasilinear evolution equations are those equations (0.1) for which the crucial assumption (K) mentioned above does not hold. It turns out that the difficulties connected with solving general strongly quasilinear evolution equations are of the same magnitude as in case of general nonlinear evolution equations.

In Chapters 7 through 9, a theory of solving general nonlinear evolution equations in nonreflexive Banach spaces is developed, and three different methods are proposed there:

1. A rapidly convergent iteration method that utilizes and further develops some of the essential technique due to Moser (1966), GLIM-III.
2. An iterative method with small steps, which is based on contractor directions, GLIM-I.
3. An iterative method with stepsize 1, GLIM-II.

All of them are based on a special kind of smooth approximate solution of the linearized equations in order to deal with the "loss of derivatives." However, what really makes the above methods work is the introduction of a new concept that is "convex approximate linearization." It should be emphasized again that Moser's (1966) method was designed for nonlinear operator equations and is not applicable to nonlinear evolution equations, although his idea was useful and very general, so that further development was possible. However, Moser's method is applicable only to strictly nonlinear operator equations, since it requires a Taylor estimate by means of a quadratic function. The same holds true for the methods of solving nonlinear evolution equations developed in

Altman (1984d; 1985a), which are based on further development of Moser's essential technique. Therefore new techniques have been proposed for solving nonlinear evolution equations under weaker differentiability conditions such as methods 2 and 3 in the above list. Moreover, method 2 has been adapted to solving quasilinear evolution equations in Altman (1984b), where the coefficients $A(\cdot,x)$ in (0.1) are locally Lipschitzian (in a certain sense), and no Taylor estimate mentioned above is available. Since the method in Altman (1984b) does not indicate how to construct smooth approximate solutions of the linearized equations (0.1), it is the purpose of this chapter to show how to do it by using, as in Chapters 2 or 8, the concept of smoothing operators in the sense of Nash (1956) and Moser (1966) combined with elliptic regularization. The case of elliptic regularization without making use of smoothing operators is also investigated.

In order to construct smooth approximate solutions of the linearized equations, one has to solve the linearized equation with a new term added, which is $\eta L(\eta)$, where L is a regularizing linear operator and $0 < \eta < 1$ is small. Now comes in the Schauder estimate for its solution and two new factors appear: the exponent $\bar{k}$, which may be called the degree of the Schauder estimate or the degree of elliptic regularization, and the norm of the solution. Here two cases are essential:

a. The solutions belong to a space with norm larger than that of the space containing the nonhomogeneous term; i.e., there is a gain of derivatives; such an estimate may be called "elliptic."
b. The solutions of the regularized-linearized equations and the nonhomogeneous terms belong to the same subspace. Such an estimate may be called a "nonelliptic" one.

It turns out that for strongly quasilinear evolution equations, it is sufficient to have available a nonelliptic Schauder estimate if smoothing operators are employed. The case without smoothing operators requires elliptic Schauder estimates. The same is true for general nonlinear evolution equations; see Chapters 8 and 9. In both cases, for strongly quasilinear evolution equations, the degree $0 < \bar{k} < 1$ of the Schauder estimate is necessary for GLIM-I. As far as the degree of the Schauder estimate is concerned the situation is different for general nonlinear evolution equations in the case involving smoothing operators, where the degree $\bar{k}$ can be slightly larger than 1. But in this case the method requires an elliptic Schauder estimate of the degree slightly larger than 1 (see pages 51, 52). However, this method is not applicable to strongly quasilinear evolution equations, as mentioned above.

1. STRONGLY QUASILINEAR EVOLUTION EQUATIONS AND CONVEX APPROXIMATE LINEARIZATION

Let $Z \subset Y \subset X$ be Banach spaces with norms $\|\cdot\|_Z \geq \|\cdot\|_Y \geq \|\cdot\|_X$. We assume that there exist positive constants C, $\bar{s}$, and p such that

$$(A_0) \quad \|x\|_Y \leq C\|x\|_X^{1-\bar{s}}\|x\|_Z^{\bar{s}} \qquad 0 < \bar{s} < 1$$

Given $0 < b$, denote by $C(0,b;X)$ the Banach space of all continuous functions $x = x(t)$ defined on the interval $[0,b]$ with values in X and norm

$$\|x\|_{\infty,X} = \sup_t[\|x(t)\|_X: \quad 0 \leq t \leq b]$$

In the same way the norms $\|y\|_{\infty,Y}$ and $\|z\|_{\infty,Z}$ are defined for Y and Z, respectively. Denote by $C^1(0,b;X)$ the vector space of all continuously differentiable functions from $[0,b]$ to X. Let W_0 be an open ball in Y with center $x_0 \in Z$ and radius $r > 0$. Put $V_0 = W_0 \cap Z$ and let V_1 be the $\|\cdot\|_Y$-closure of V_0. Corresponding to each $(t,x) \in [0,b] \times V_1$ let the mapping $A(t,x)$: $Y \to X$ be linear and continuous. Consider the Cauchy problem for the quasilinear equation

$$Px(t) \equiv \frac{dx}{dt} + A(t,x)x - f(t,x) = 0 \tag{1.1}$$

$0 \leq t \leq b$, $x(0) = x_0$. Let G be the set of functions

$$x \in C(0,b;V_0(\|\cdot\|_Z))C^1(0,b;X) \qquad \text{with } x(0) = x_0$$

1.1 DEFINITION Let μ, ν, and σ be positive constants with $\nu < 1$. Then the linearized equations

$$\frac{dz}{dt} + A(t,x)z - f(t,x) = 0 \qquad 0 \le t \le b \qquad z(0) = x_0 \tag{1.2}$$

admit smooth approximate solutions of order (μ,ν,σ) if there exists a constant $M > 0$ with the following property. For every $x \in G$, $K > 1$, and $Q > 1$ if $\|x\|_{\infty,Z} < K$, then there exist a residual function y and a function z such that

$$\|y\|_{\infty,X} \le MQ^{-\mu}K^{\sigma} \qquad \sigma,\mu > 0 \tag{1.3}$$

$$\|z\|_{\infty,Z} \le MQK^{\nu} \qquad 0 < \nu < 1 \tag{1.4}$$

and

$$\frac{dz}{dt} + A(t,x)z - f(t,x) + y = 0 \qquad 0 \le t \le b \qquad z(0) = x_0 \tag{1.5}$$

This definition generalizes the concept of smooth approximate solutions of linearized equations introduced in Altman (1984d) and is also related to the one introduced by Moser (1966) for operator equations in case of $\sigma = \nu = 1$.

Let us mention that the case $\nu = 1$ is not accessible to the method proposed here or in Altman (1984d). We make the following assumptions.

(A_1) $A(\cdot,x)$ and $f(\cdot,x)$ are uniformly continuous in the following sense. $x_n \in G$ and $\|x_n - x\|_{\infty,Y} \to 0$ imply

$$\|A(\cdot,x_n) - A(\cdot,x)\|_{\infty,X} \to 0 \qquad \text{as } n \to \infty \tag{1.6}$$

$$\|f(\cdot,x_n) - f(\cdot,x)\|_{\infty,X} \to 0 \qquad \text{as } n \to \infty \tag{1.7}$$

(A_2) There exists a constant L such that

$$\|[A(t, v + \varepsilon h) - A(t,v)]u\|_X \le \varepsilon L\|h\|_X\|u\|_Y \tag{1.8}$$

where $v + \varepsilon h \in V_0$, and

$$\|f(t, v + \varepsilon h) = f(t,v)\|_X \le \varepsilon\|h\|_X \tag{1.9}$$

for all $(t,v) \in [0,b] \times V_1$ and $h \in Y$.

(A_3) There exists a constant $C > 0$ with the following property. For $x \in G$ with $\|x - x_0\|_{\infty,Y} < r$ and g with $\|g\|_{\infty,X} < \infty$, if h is a solution of the equation

$$\frac{dh}{dt} + A(t,x)h + g = 0 \qquad 0 \le t \le b \qquad h(0) = 0 \tag{1.10}$$

then

$$\|h\|_{\infty,X} \le bC\|g\|_{\infty,X} \tag{1.11}$$

and the linearized equation (1.2) admits smooth approximate solutions in the sense of Definition 1.1.

(A_4) We assume that

$$\|A(\cdot,x_0)x_0 - f(\cdot,x_0)\|_{\infty,X} \le p_0 \tag{1.12}$$

Now for $x \in G$ let z be a solution of the equation (1.5) and put $z = x + h$. Then obviously h is a solution of the equation

$$\frac{dh}{dt} + A(t,x)h + Px + y = 0 \qquad 0 \le t \le b \qquad h(0) = 0 \tag{1.13}$$

and we have

$$\|h\|_{\infty,X} \le bC(\|Px\|_{\infty,X} + MQ^{-\mu}K^{\sigma}) \tag{1.14}$$

and

$$\|h\|_{\infty,Z} \le \|x\|_{\infty,Z} + MQK^{\nu} \tag{1.15}$$

by virtue of (1.3), (1.4), and (1.11) with g replaced by $Px + y$.

1.1 LEMMA Let $x \in G$ with $\|x - x_0\|_{\infty,Y} < r$, and let $0 < \bar{q} < q < 1$ be arbitrary fixed numbers. If h is a solution of the equation (1.13), then there exists $0 < \varepsilon \le 1$ such that

$$\|P(x + \varepsilon h) - (1 - \varepsilon)Px\|_{\infty,X} \le \varepsilon q\|Px\|_{\infty,X} \tag{1.16}$$

and

$$\|h\|_{\infty,X} \le bC(1 + \bar{q})\|Px\|_{\infty,X} \tag{1.17}$$

Proof: We have the following identity:

$$P(x + \varepsilon h) - (1 - \varepsilon)Px = [A(\cdot, x + \varepsilon h) - A(\cdot,x)](x + \varepsilon h) - [f(\cdot, x + \varepsilon h) - f(\cdot,x)] + \varepsilon[dh/dt + A(t,x)h + Px + y] - \varepsilon y$$

To obtain an estimate for the first two terms of the above identity we use relations (1.8) and (1.9) with small ε. The third term is equal to zero. Hence, we obtain by (1.3) and (1.14),

$$\begin{aligned}\|P(x + \varepsilon h) - (1 - \varepsilon)Px\|_{\infty,X} &\leq \varepsilon L(1 + r)\|h\|_{\infty,X} + \varepsilon MQ^{-\mu}K^{\sigma} \\ &\leq \varepsilon L(1 + r)bC\|Px\|_{\infty,X} + \varepsilon 2MQ^{-\mu}K^{\sigma} \\ &\leq \varepsilon(q - \bar{q})\|Px\|_{\infty,X} + \varepsilon 2MQ^{-\mu}K^{\sigma} \\ &\leq \varepsilon q\|Px\|_{\infty,X}\end{aligned}$$

provided b is so small that

$$L(1 + r)bC < q - \bar{q} < 1 \quad \text{and} \quad bC < 1 \tag{1.18}$$

and the following relation holds which also yields (1.17) by (1.14).

$$2MQ^{-\mu}K^{\sigma} \leq \bar{q}\|Px\|_{\infty,X} \quad \blacksquare \tag{1.19}$$

1.2 LEMMA Let $x_n \in G$ be such that $\|x_n - x_0\|_{\infty,Y} < r$, and let

$$\|x_n\|_{\infty,Z} < A \exp(\alpha(1 - q)t_n) = K_n \tag{1.20}$$

where $t_n > 0$,

$$\mu(1 - \nu) - \sigma > 0 \tag{1.21}$$

where α and A are chosen so that

$$\alpha(1 - q) - 1 = \delta > 0 \quad \text{and} \quad \alpha > (\mu(1 - \nu) - \sigma)^{-1} \tag{1.22}$$

and

$$M(2M)^{1/\mu}(\bar{q}p_0)^{-1/\mu} < A^c(\alpha(1 - q) - 1) \tag{1.23}$$

where $c = 1 - \nu - \sigma/\mu$. Then there exists a number $Q = Q_n$ such that

$$2MQ_n^{-\mu}K_n^{\sigma} < \bar{q}p_0 \exp(-(1 - q)t_n) \tag{1.24}$$

and

$$\|x_n + \varepsilon h_n\|_{\infty,Z} < A \exp(\alpha(1 - q)(t_n + \varepsilon)) \tag{1.25}$$

where h_n is a solution of the equation (1.13) with $x = x_n$ which satisfies relation (1.15).

Proof: Since, by virtue of (1.15), we have

$$\begin{aligned}\|x_n + \varepsilon h_n\|_{\infty,Z} &\leq \|x_n\|_{\infty,Z} + \varepsilon\|h_n\|_{\infty,Z} \\ &\leq (1 + \varepsilon)\|x_n\|_{\infty,Z} + \varepsilon MQ_nK_n^{\nu}\end{aligned}$$

relation (1.25) is satisfied if

$$(1 + \varepsilon)A \exp(\alpha(1 - q)t_n) + \varepsilon MQK^{\nu} < A \exp(\alpha(1 - q)(t_n + \varepsilon)) \tag{1.26}$$

But it is easily seen that

$$\delta = \alpha(1 - q) - 1 = \min_{0\leq\varepsilon\leq1} [\exp(\alpha(1 - q)\varepsilon) - (1 + \varepsilon)]/\varepsilon \tag{1.26a}$$

Hence, by (1.26a), $Q = Q_n$ is subject to

$$Q < (\alpha(1 - q) - 1)A^{1-\nu}M^{-1} \exp(\alpha(1 - \nu)(1 - q)t_n) \tag{1.27}$$

On the other hand, (1.24) yields

$$(2M)^{1/\mu}(\bar{q}p_0)^{-1/\mu}A^{\sigma/\mu} \exp(\sigma\alpha + 1)(1 - q)t_n/\mu) < Q \tag{1.28}$$

Hence, it follows that relations (1.27) and (1.28) are satisfied if conditions (1.21) to (1.23) hold. This completes the proof of the lemma. ∎

Now we shall construct an iterative method of contractor directions as follows. Put $x_0 = 0$ and assume that $x_0, x_1, \ldots, x_n \in G$ and $t_0 = 0, t_1, \ldots, t_n > 0$ are known and satisfy the following relations:

$$\|x_n - x_0\|_{\infty,Y} < r \tag{1.29}$$

$$\|x_n\|_{\infty,Z} < A \exp(\alpha(1 - q)t_n) \tag{1.30}$$

$$\|Px_n\|_{\infty,X} \le p_0 \exp(-(1 - q)t_n) \tag{1.31}$$

where α and A are subject to (1.21) to (1.23). Then put $t_{n+1} = t_n + \varepsilon_n$, and

$$x_{n+1} = x_n + \varepsilon_n h_n \tag{1.32}$$

That is, we have the convex iteration

$$x_{n+1} = (1 - \varepsilon_n)x_n + \varepsilon_n z_n$$

where $0 < \varepsilon_n \le 1$ is such that (1.29) holds for n + 1, and h_n and $z_n = x_n + h_n$ are solutions of the equations (1.13) and (1.5), respectively, with $x = x_n$; and by Lemma 1.2, $Q = Q_n$ can be such that relation (1.25) holds for ε_n, h_n yielding (1.30) for n + 1 as well as

$$\begin{aligned} &\|P(x_n + \varepsilon_n h_n) - (1 - \varepsilon_n)Px_n\|_{\infty,X} \\ &\le \varepsilon_n q p_0 \exp(-(1 - q)t_n) \end{aligned} \tag{1.33}$$

by virtue of (1.16), where $x = x_n$, $h = h_n$, $\varepsilon = \varepsilon_n$, and $\|Px_n\|_{\infty,X}$ is replaced by its estimate

$$\|Px_n\|_{\infty,X} \le p_0 \exp(-(1 - q)t_n) \tag{1.33a}$$

obtained by induction. It results from (1.33) that

$$\begin{aligned} \|Px_{n+1}\|_{\infty,X} &\le (1 - (1 - q)\varepsilon_n)\|Px_n\|_{\infty,X} \\ &\le \exp(-(1 - q)\varepsilon_n)\|Px_n\|_{\infty,X} \\ &\le \|Px_0\|_{\infty,X} \exp(-(1 - q)(t_n + \varepsilon_n)) \end{aligned}$$

Hence, relation (1.31) holds for n + 1. By the same argument we also get from (1.19), (1.17) with $h = h_n$, $x = x_n$, and $\|Px_n\|_{\infty,X}$ replaced by its estimate (1.31),

$$\|h_n\|_{\infty,X} \le 2bCp_0 \exp(-(1 - q)t_n) \tag{1.34}$$

and

$$\|h_n\|_{\infty,Z} \le (1 + \delta)A \exp(\alpha(1 - q)t_n) \tag{1.35}$$

Relation (1.35) results from (1.15) with $h = h_n$, $x = x_n$, $Q = Q_n$, and $K = K_n$ from (1.20), (1.26), and (1.26a).

Relations (1.34), (1.35), and (A_0) imply

$$\|h_n\|_{\infty,Y} \le CNb^{1-\bar{s}} \exp(-(1 - q)(1 - \bar{s}(1 + \alpha))t_n) \tag{1.36}$$

where

$$\begin{aligned} N &= (2Cp_0)^{1-\bar{s}} [(1 + \delta)A]^{\bar{s}} \\ &= (2Cp_0)^{1-\bar{s}} [\alpha(1 - q)A]^{\bar{s}} \end{aligned}$$

provided

$$1 - \bar{s}(1 + \alpha) = \lambda > 0 \tag{1.37}$$

1.3 LEMMA The following estimate results from (1.36).

$$\sum_{n=0}^{\infty} \varepsilon_n \|h_n\|_{\infty,Y} \le CNb^{1-s} \exp((1 - q)\lambda) [1 - q)\lambda]^{-1} \tag{1.38}$$

Proof: We have by (1.36) and (1.37),

$$\begin{aligned} &\sum_{n=0}^{\infty} \varepsilon_n \exp(-(1 - q)\lambda t_n) \\ &= \sum_{n=0}^{\infty} (t_{n+1} - t_n) \exp(-(1 - q)\lambda t_n) \\ &= \sum_{n=0} (t_{n+1} - t_n) \exp(-(1 - q)\lambda t_{n+1}) \exp((1 - q)\lambda \varepsilon_n) \\ &\le \exp((1 - q)\lambda) \sum_{n=0}^{\infty} \int_{t_n}^{t_{n+1}} \exp(-(1 - q)\lambda t)\, dt \\ &= \exp((1 - q)\lambda) \int_0^{\infty} \exp(-(1 - q)\lambda t)\, dt \\ &= \exp((1 - q)\lambda)[(1 - q)\lambda]^{-1} \end{aligned}$$

Suppose that

$$CNb^{1-\bar{s}} \exp((1 - q)\lambda)[(1 - q)\lambda]^{-1} < r \tag{1.39}$$

where N is determined by (1.36). Then it follows from (1.36) that there exists x such that

$$\|x_n - x\|_{\infty,Y} \to 0 \qquad \text{and} \qquad \|x - x_0\|_{\infty,Y} < r \tag{1.40}$$

Moreover, relations (1.33) and (1.33a) imply that

$$\|Px_{n+1} - Px_n\|_{\infty,X} \le \varepsilon_n(1 + q)p_0 \exp(-(1 - q)t_n)$$

Hence, using the same argument as in Lemma 1.3, we can see that $\{Px_n\}$ is a Cauchy sequence, and

$$\|Px_n - Px\|_{\infty,X} \to 0 \qquad \text{as } n \to \infty \tag{1.41}$$

by virtue of (1.6) and (1.7). ∎

1.4 LEMMA Suppose that condition (1.39) is satisfied and $\{x_n\} \subset G$ is an arbitrary sequence which satisfies relations (1.40), (1.41), and (1.29) to (1.31), where $\lim_{n\to\infty} t_n = t < \infty$. Then there exist $0 < \varepsilon \le 1$ and $\bar{x}$ such that

$$\|P\bar{x} - (1 - \varepsilon)Px\|_{\infty,X} \le \varepsilon q' p_0 \exp(-(1 - q)t) \tag{1.42}$$

where $\bar{x} \in G$, $\|\bar{x} - x_0\|_{\infty,Y} < r$, $\bar{q} < q' < q < 1$, and $C < C_1 < \bar{C}_1$, and

$$\|\bar{x} - x\|_{\infty,Y} \le \varepsilon C_1 N_1 b^{1-s} \exp(-(1 - q)\lambda t) \tag{1.43}$$

with N, s, and λ as in (1.36), (1.37), and N_1 is N with C replaced by C_1.

Proof: Choose $0 < \varepsilon \le 1$ such that

$$\varepsilon C_1 N_1 b^{1-s} \exp(-(1 - q)\lambda t) < r - \|x - x_0\|_{\infty,Y}$$

and n so large that

$$\|x_n - x\|_{\infty,Y} < r - \|x - x_0\|_{\infty,Y} - \varepsilon C_1 N_1 b^{1-s/p} \exp(-(1 - q)\lambda t) \tag{1.44}$$

Having chosen ε, we then take x_n, Px_n closer to x, Px, respectively, so as to satisfy the following relations.

$$\|Px_n - Px\|_{\infty,X} \le \varepsilon(q' - \bar{q})2^{-1}p_0 \exp(-(1 - q)t) \tag{1.45}$$

$$\|x_n - x\|_{\infty,Y} \le \varepsilon(C_1 - C)N_1 b^{1-\bar{s}} \exp(-(1 - q)t) \tag{1.46}$$

with N_1 as in (1.43).

Let $z_n = x_n + h_n$ and h_n be solutions of the equations (1.5) and (1.13), respectively, with $x = x_n$, $h = h_n$, and $0 < \bar{q} < q' < q < 1$. Thus, we have

$$\frac{dz_n}{dt} + A(t,x_n)z_n - f(t,x_n) + y = 0 \qquad 0 \le t \le b \qquad z(0) = x_0$$

and

$$\frac{dh_n}{dt} + A(t,x_n)h_n = Px_n + y = 0 \qquad 0 \le t \le b \qquad h_n(0) = 0$$

We get from (1.14) with $x = x_n$ and $h = h_n$,

$$\begin{aligned}\|h_n\|_{\infty,X} &\le bC(\|Px_n\|_{\infty,X} + MQ^{-\mu}K^{\sigma}) \\ &\le bC(\|Px\|_{\infty,X} + \|Px - Px_n\|_{\infty,X} + MQ^{-\mu}K^{\sigma}) \\ &\le bC_1(\|Px\|_{\infty,X} + MQ^{-\mu}K^{\sigma}) \\ &\le bC_1(1 + \bar{q})\|Px\|_{\infty,X}\end{aligned}$$

$$\|h_n\|_{\infty,X} \le bC_1(1 + \bar{q})p_0 \exp(-(1 - q)t) \tag{1.47}$$

provided that (1.19) holds with the right-hand term replaced by its estimate, i.e., if

$$2MQ^{-\mu}K^{\sigma} \le \bar{q}p_0 \exp(-(1 - q)t) \tag{1.48}$$

and if (1.44) holds, since

$$\|x_n\|_{\infty,Z} < A \exp(\alpha(1 - q)t) = K \tag{1.49}$$

by virtue of (1.30), since $t_n \uparrow t$ as $n \to \infty$. We also get from (1.15), by (1.4) and (1.49),

$$\begin{aligned}\|h_n\|_{\infty,Z} &\le \|x_n\|_{\infty,Z} + \|z_n\|_{\infty,Z} \\ &< A \exp(\alpha(1 - q)t) + MQK^{\nu} < (1 + \delta)A \exp(\alpha(1 - q)t)\end{aligned}$$

with δ as in (1.26a), where Q satisfies (1.27), and (1.28) with t_n replaced by t. Hence,

$$\|h_n\|_{\infty,Z} < (1 + \delta)A \exp(\alpha(1 - q)t) \tag{1.50}$$

Relations (1.47), (1.50), and (A_0) imply

$$\|h_n\|_{\infty,Y} < CN_1 b^{1-\bar{s}} \exp(-(1 - q)\lambda t) \tag{1.51}$$

with $\bar{s}$, λ, and N_1 as in (1.36) and (1.37), where C in N is replaced by $C_1 > C$ and $b\bar{C} < 1$.

In order to prove relation (1.42), we use the following identity for $\bar{x} = x_n + \varepsilon h_n$.

$$\begin{aligned} P\bar{x} - (1 - \varepsilon)Px &= P(x_n + \varepsilon h_n) - (1 - \varepsilon)Px_n + (1 - \varepsilon)(Px_n - Px) \\ &= [A(\cdot, x_n + \varepsilon h_n) - A(\cdot, x_n)](x_n + \varepsilon h_n) \\ &\quad - [f(\cdot, x_n + \varepsilon h_n) - f(\cdot, x_n)] \\ &\quad + \varepsilon[dh_n/dt + A(t, x_n)h_n + Px_n + y] \\ &\quad - \varepsilon y + (1 - \varepsilon)(Px_n - Px) \end{aligned}$$

To obtain an estimate for the first two terms we use relations (1.8) and (1.9) with sufficiently small $0 < \varepsilon \le 1$. The third term is equal to zero. Hence we obtain, by (1.3) and (1.14) with $x = x_n$ and $h = h_n$,

$$\begin{aligned} \|P\bar{x} - (1 - \varepsilon)Px\|_{\infty,X} &\le \varepsilon 2Lr\|h_n\|_{\infty,X} + \varepsilon MQ^{-\mu}K^{\sigma} \\ &\quad + \varepsilon(q' - \bar{q})2^{-1}p_0 \exp(-(1 - q)t) \\ &\le \varepsilon 2LrbC_1(1 + \bar{q})p_0 \exp(-(1 - q)t) \\ &\quad + \varepsilon(q' - \bar{q})2^{-1}p_0 \exp(-(1 - q)t) \\ &\quad + \varepsilon\bar{q}p_0 \exp(-(1 - q)t) \end{aligned}$$

by virtue of (1.46), (1.45), (1.47), and (1.48). Hence, relation (1.42) holds provided

$$2LrbC_1(1 + \bar{q}) \le \frac{q' - \bar{q}}{2} \tag{1.52}$$

Note that if b satisfies (1.52) then the weaker condition (1.18) holds true. Now we have for $\bar{x} = x_n + h_n$, where $n = \bar{n}$ is sufficiently large,

$$\|\bar{x} - x_0\|_{\infty,Y} \le \|x_n - x\|_{\infty,Y} + \varepsilon\|h_n\|_{\infty,Y} + \|x - x_0\|_{\infty,Y} < r$$

by virtue of (1.44) and (1.51). Finally, let $\bar{C}$ be the same as in (1.43). Then we obtain for $\bar{x} = x_n + \varepsilon h_n$, $n = \bar{n}$,

$$\|\bar{x} - x\|_{\infty,Y} \le \|x_n - x\|_{\infty,Y} + \varepsilon\|h_n\|_{\infty,Y}$$
$$\le \varepsilon[(C_1 - C) + C]N_1 b^{1-\bar{s}} \exp(-(1 - q)t)$$

by virtue of (1.46) and (1.51). Hence, relation (1.43) follows. ∎

We are now in a position to prove the following.

1.1 THEOREM Suppose that in addition to the hypotheses (A_0) to (A_4), conditions (1.21) to (1.23) are satisfied and $0 < b' \le b$ is such that $b'\bar{C} < 1$ with $C < \bar{C}$, where C is from (1.11) and (A_0), and

$$\bar{C}N_1 b'^{1-\bar{s}} \exp((1 - q)\lambda)[(1 - q)\lambda]^{-1} < r$$

Then the equation (1.1) with b replaced by b' has a solution x, and there exists a sequence $\{x_n\} \subset G$ such that

$$\|x_n - x\|_{\infty,Y} \to 0 \qquad \text{as } n \to \infty$$

$$\|x_n - x_0\|_{\infty,Y} < r \qquad \text{and} \qquad \|x - x_0\|_{\infty,Y} < r$$

Proof: Although the iterative sequence determined by (1.32) is convergent, its limit function may not be a solution. Therefore we replace that sequence by an optimal one in the following sense. Suppose that $x_0(t) \equiv x_0$, and $x_1, x_2, \ldots, x_n$ are known. Then relations (1.33) and (1.36) imply that there exist $\bar{x}_n$ and $0 < \bar{\varepsilon}_n \le 1$ such that

$$\|P\bar{x}_n - (1 - \bar{\varepsilon}_n)Px_n\|_{\infty,X} \le \bar{\varepsilon}_n q p_0 \exp(-(1 - q)t_n) \qquad (1.53)$$

and

$$\|\bar{x}_n - x_n\|_{\infty,Y} \le \bar{\varepsilon}_n \bar{C} N_1 b'^{1-\bar{s}} \exp(-(1 - q)\lambda t_n) \tag{1.54}$$

Considering the set of all such $(\bar{x}_n, \bar{\varepsilon}_n)$ we choose $x_{n+1} = \bar{x}_n$ with approximately the largest possible $\varepsilon_n = \bar{\varepsilon}_n$, e.g., $\varepsilon_n > c \sup \bar{\varepsilon}_n$ with $0 < c < 1$ so that relations (1.53) and (1.54) hold for $\bar{x}_n = x_{n+1}$ and $\bar{\varepsilon}_n = \varepsilon_n$. Now it is easily seen that the sequence $\{x_n\}$ just constructed satisfies the hypotheses of Lemma 1.4. Then it results from (1.42) that the relation

$$\|P\bar{x} - (1 - \varepsilon)Px_n\|_{\infty,X} \le \varepsilon q p_0 \exp(-(1 - q)t_n) \tag{1.55}$$

is satisfied for almost all n; because $q' < q$, the sequence $\{Px_n\}$ is convergent to Px, and t is the limit of the increasing sequence $\{t_n\}$. By the same argument it follows from (1.43) that the relation

$$\|\bar{x} - x_n\|_{\infty,Y} \le \varepsilon \bar{C} N_1 b^{1-s/p} \exp(-(1 - q)\lambda t_n)$$

holds for almost all n, since $C_1 < \bar{C}$ and $\{x_n\}$ converges to x. But the maximal choice of ε_n implies that $\varepsilon_n \ge c\varepsilon$ for almost all n. Since $t_{n+1} = t_n + \varepsilon_n$, the sequence $\{t_n\}$ converges to ∞. Hence, by virtue of (1.33a), we get

$$\|Px_n\|_{\infty,X} \to 0 \qquad \text{as } n \to \infty$$

This completes the proof of the theorem. ∎

1.1 REMARK Theorem 1.1 yields only existence of a solution of the quasilinear evolution equation. However, it is easily seen that a uniqueness theorem can be obtained if we assume the following stronger Lipschitz conditions:

$$\|[A(t,u) - A(t,v)]y\|_X \le K_1\|u - v\|_X\|y\|_Y$$

and

$$\|f(t,u) - f(t,v)\|_X \le K_2\|u - v\|_Y$$

for all $t \in [0,b]$, $u,v \in V_1$, and $y \in Y$ and some $K_1, K_2 > 0$.

In fact, suppose that $\bar{x}$ and x are solutions of equation (1.1); then we get

$$\frac{d(\bar{x} - x)}{dt} + A(t,\bar{x})(\bar{x} - x) = f(t,\bar{x}) - f(t,x) - [A(t,\bar{x}) - A(t,x)]\bar{x}$$

$0 \le t \le b'$, $\bar{x}(0) - x(0) = 0$. Hence, the uniqueness follows if b' is small.

As in [5] one can show that approximate solutions in the sense of Definition 1.1 can be obtained by using smoothing operators.

An Application of the Theory of C_0-Semigroups

Sufficient conditions for the existence of a constant C in (A_3) can be obtained by making use of the theory of C_0-semigroups.

Denote by $G(X)$ the set of all negative infinitesimal generators of C_0-semigroups $\{U(t)\}$ on X. If $A \in G(X)$, then $\{U(t)\} = \{e^{-tA}\}$, $0 \le t < \infty$, is the semigroup generated by $-A$. For $x \in G$, put $A(t) = A(t,x(t))$. We assume that A is a function from $[0,b] \times V_0$ into $G(X)$, and $A(t)$ is stable in X (see Kato, 1970, 1973, 1975; Yosida, 1968; Tanabe, 1979). We also assume that the evolution operator $U(t,s;x)$ generated by $A(t,x(t))$ exists and satisfies the following condition:

$$\|U(t,s;x)\|_{X\to X} \le M_X$$

for all $x \in G$ and some constant $M_X > 0$. Note that evolution operators $U(t,s;x)$ can be replaced by their approximations $U_m(t,s;x)$ generated by the appropriate step functions (see Kato, 1970, 1973, 1975).

2. SMOOTHING OPERATORS COMBINED WITH ELLIPTIC REGULARIZATION FOR STRONGLY QUASILINEAR EVOLUTION EQUATIONS AND AN ITERATIVE METHOD OF CONTRACTOR DIRECTIONS

Let $\{X_j\}$ with $0 \le j \le p$ be a scale of Banach spaces with increasing norms such that $i < j$ implies $x_j \subset X_i$ and $\|x\|_j \ge \|x\|_i$, and let $0 < m_1 < m_2 < \bar{p} < p$; $j = m_1, m_2, \bar{p}, p$, and some s.

We make the following assumptions.

(A_0) We assume that there exists a family of linear operators S_θ, $\theta \geq 1$ [smoothing operators in the sense of Nash (1956) and Moser (1966)] such that

$$\|(I - S_\theta)x\|_0 \leq C\theta^{-m_1}\|x\|_{m_1}$$

$$\|(I - S_\theta)x\|_0 \leq C\theta^{-(m_2-m_1)}\|x\|_{m_2}$$

$$\|S_\theta x\|_p \leq C\theta^{p-i}\|x\|_i \qquad i = m_2 \text{ or } \bar{p}$$

for some constant $C > 0$, where I is the identity mapping. We also assume that $\|x\|_j \leq C\|x\|_i^{1-\beta}\|x\|_p^{\beta}$, for $0 \leq \beta \leq 1$.

Given $0 < b$, denote by $C(0,b;X_j)$ the Banach space of all continuous functions $x = x(t)$ defined on the interval $[0,b]$ with values in X_j and norms $\|x\|_{\infty,j} = \sup_t [\|x(t)\|_j\colon\ 0 \leq t \leq b]$, for $j = 0, m_1, m_2, p$ and s to be determined later. Denote by $C^1(0,b;X_0)$ the vector space of all continuously differentiable functions from $[0,b]$ to X_0. Let $W_0 \subset X_s$ be an open ball with center x_0 and radius $r > 0$. Put $V_0 = W_0 \cap X_p$ and let V_s be the $\|\cdot\|_s$-closure of V_0.

Let G be a set of functions $x \in C(0,b;V_0(\|\cdot\|_p)) \cap C^1(0,b;X_0)$ with $x(0) = x_0$, $\|x\|_{\infty,s} < r$, and $\|x\|_{\infty,p} < \infty$.

Corresponding to each given $(t,x) \in [0,b] \times V_s$, let the mapping $A(t,x)\colon\ X_s \to X_0$ be linear and continuous. Consider the Cauchy problem for the quasilinear equation

$$Px(t) \equiv \frac{dx}{dt} + A(t,x)x - f(t,x) = 0 \tag{2.1}$$

$$0 \leq t \leq b \qquad x(0) = x_0 = 0$$

2.1 DEFINITION Let μ, ν, and σ be positive numbers with $\nu < 1$. Then the linearized equation

$$\frac{dz}{dt} + A(t,x)z - f(t,x) = 0 \qquad 0 \leq t \leq b \qquad z(0) = x_0 \tag{2.2}$$

admits smooth approximate solutions of order (μ,ν,σ) if there exists a constant M which has the following property. For every $x \in G$, $K > 1$, and $Q > 1$ if $\|x\|_{\infty,Z} < K$, then there exist a residual (error) function y and a function z such that

$$\|z\|_{\infty,p} < MQK^{\nu} \qquad 0 < \nu < 1 \tag{2.3}$$

and

$$\|y\|_{\infty,0} \le MQ^{-\mu}K^{\sigma} \tag{2.4}$$

and such that

$$\frac{dz}{dt} + A(t,x)z - f(t,x) + y = 0 \qquad 0 \le t \le b \qquad z(0) = x_0 \tag{2.5}$$

We make the following assumptions.

(A_1) $A(\cdot,x)$ and $f(\cdot,x)$ are uniformly continuous in the following sense. $\{x_n\} \subset G$ and $\|x_n - x\|_{\infty,s} \to 0$ imply

$$\|A(\cdot,x_n) - A(\cdot,x)\|_{\infty,X_s \to X_0} \to 0 \tag{2.6}$$

$$\|f(\cdot,x_n) - f(\cdot,x)\|_{\infty,X_0} \to 0 \qquad \text{as } n \to \infty \tag{2.7}$$

(A_2) There exists a constant C such that

$$\|A(t, v + \varepsilon h) - A(t,v)]u\|_0 \le C\varepsilon\|h\|_0\|u\|_s \tag{2.8}$$

where $v + \varepsilon h \in V_0$; $(t,v) \in [0,b] \times V_1$,

$$\|A(t,x)h\|_{m_1} \le C\|h\|_{m_2} \tag{2.9}$$

where $(t,x) \in [0,b] \times V_0$,

$$\|A(t,x)h\|_0 \le C\|h\|_{m_1} \tag{2.10}$$

where $(t,x) \in [0,b] \times V_0$.

(A_3) There exists a constant $C > 0$ such that

$$\|f(t, v + \varepsilon h) - f(t,v)\|_0 \le C\varepsilon\|h\|_0 \tag{2.11}$$

for all $(t,v) \in [0,b] \times V_1$, and $v + \varepsilon h \in V_1$. We also assume that

$f(\cdot,x)$ can be approximated as follows. If $x \in G$, then for arbitrary $\omega > 0$, there exists a function g_ω such that $g_\omega \in C(0,b;X_p(\|\cdot\|_p))$ and

$$\|f(\cdot,x) - g_\omega\|_{\infty,m_2} \le \omega \tag{2.12}$$

In addition, there exist d_0 and p_0 such that

$$\|f(\cdot,x)\|_{\infty,m_2} \le d_0 \qquad x \in G \tag{2.13}$$

and

$$\|A(\cdot,x_0)x_0 - f(\cdot,x_0)\|_{\infty,0} \le p_0 \tag{2.14}$$

(A_4) There exists a linear (regularizing) operator $L = L(\eta)$ such that

$$\|Lx\|_{m_1} \le C\|x\|_{m_2} \tag{2.15}$$

for some constant $C > 0$.

There exists $0 < \bar{k} < 1$ such that the linearized equation

$$\frac{dz}{dt} + A(t,x)z + \eta Lz - g_\omega = 0 \qquad 0 \le t \le b \qquad z(0) = x_0 \tag{2.16}$$

with small $\eta > 0$, has a solution $\bar{z}$ such that

$$\|\bar{z}\|_{\infty,i} \le C\eta^{-\bar{k}}\|g_\omega\|_{\infty,m_2} \qquad i = m_2 \text{ or } \bar{p} \tag{2.17}$$

for some $C > 0$.

(A_5) For $x \in G$ and g with $\|g\|_{\infty,0} < \infty$, if h is a solution of the equation

$$\frac{dh}{dt} + A(t,x)h + g = 0 \qquad 0 \le t \le b \qquad h(0) = 0 \tag{2.18}$$

then

$$\|h\|_{\infty,0} \le bC\|g\|_{\infty,0} \tag{2.19}$$

for some constant $C > 0$.

From now on in this section $x_0 = 0$. For $x \in G$, let $\bar{z}$ be a solution of the equation (2.16) satisfying (2.17). Then the following estimates hold.

2.1 LEMMA The following relations hold for $x \in G$ and η such that $0 < \eta < 1$.

$$\|(I - S_\theta)\frac{d\bar{z}}{dt}\|_{\infty,0} \le M_1\theta^{-m_1}\eta^{-k} \tag{2.20}$$

$$\|A(\cdot,x)(I - S_\theta)\bar{z}\|_{\infty,0} \le M_2\theta^{-(m_2-m_1)}\eta^{-k} \tag{2.21}$$

$$\|z\|_{\infty,p} = \|S_\theta\bar{z}\|_{\infty,p} \le M_3\theta^{p-i}\eta^{-k} \tag{2.22}$$

for some constants M_1, M_2, M_3 and $z = S_\theta\bar{z}$.

Proof: We get from (2.16) and (2.17),

$$\begin{aligned}\|(I - S_\theta)\frac{d\bar{z}}{dt}\|_0 &= \|(I - S_\theta)[A(t,x)\bar{z}(t) + \eta L\bar{z}(t) - g_\omega(t)\|_0\\ &\le C\theta^{m_1}(\|A(t,x)\bar{z}(t)\|_{m_1} + \eta C\|\bar{z}(t)\|_{m_2} + \|g_\omega(t)\|_{m_1})\\ &\le C\theta^{-m_1}(2C\|\bar{z}(t)\|_{m_2} + \|g_\omega(t)\|_{m_2})\\ &\le C\theta^{-m_1}2C_1(\eta^{-k} + 1)\|g_\omega\|_{\infty,m_2}\end{aligned}$$

Hence, relation (1.20) follows, by virtue of (2.12), (2.13), and (2.15), since

$$\|\bar{z}\|_{\infty,m_2} \le C_2\eta^{-k} \tag{2.23}$$

and

$$\|g_\omega\|_{\infty,m_2} \le d_1$$

for some constants C_2, d_1. We have

$$\|A(t,x(t))(I - S_\theta)\bar{z}\|_0 \le C\|(I - S_\theta)\bar{z}(t)\|_{m_1} \le C\theta^{-(m_2-m_1)}\|\bar{z}\|_{\infty,m_2}$$

and relation (2.21) results from (2.23). Finally, we get from (A_0)

$$\|S_\theta\bar{z}(t)\|_p \le C\theta^{-(p-i)}\|\bar{z}\|_{\infty,i}$$

and (2.22) follows from (2.23).

Consider the linearized equation

$$\frac{dz}{dt} + A(t,x)z - f(t,x) = 0 \qquad 0 \le t \le b \qquad z(0) = 0 \tag{2.24}$$

where $x \in G$, and put $z = S_\theta \bar{z}$, $\bar{z}$ being a solution of the linearized equation (2.16) with $x_0 = 0$ which satisfies condition (2.17). ∎

2.2 LEMMA The following estimate holds.

$$\left\|\frac{dz}{dt} + A(t,x)z - f(t,x)\right\|_{\infty,0} \le M_4(\theta^{-m} + \eta)\eta^{-k} + \omega \tag{2.25}$$

for $x \in G$ and some constant $M_4 > 0$, where $\omega > 0$ is arbitrary, and

$$m = \min(m_1, m_2 - m_1) \tag{2.26}$$

Proof: We have

$$\begin{aligned}
&\left\|\frac{dz}{dt} + A(t,x)z - f(t,x)\right\|_{\infty,0} \\
&\le \left\|\frac{d\bar{z}}{dt} + A(t,x)\bar{z} + \varepsilon L\bar{z} - g_\omega\right\|_{\infty,0} + \|f(\cdot,x) - g_\omega\|_{\infty,0} \\
&\quad + \left\|(I - S_\theta)\frac{d\bar{z}}{dt}\right\|_{\infty,0} + \|A(\cdot,x)(I - S_\theta)\bar{z}\|_{\infty,0} + \eta\|L\bar{z}\|_{\infty,0}
\end{aligned}$$

and relation (2.25) results from Lemma 2.1, (2.15), (2.16), (2.17), and (2.12).

Now put

$$Q = \theta^{p-\bar{p}} \tag{2.27}$$

$$\mu = \frac{m}{p - \bar{p}} \tag{2.28}$$

Let $0 < \nu < 1$ and choose $0 < \eta < 1$ so as to satisfy the relations

$$\eta^{-k} < K^{\nu} \qquad \text{and} \qquad \eta < Q^{-\mu} \qquad (K > 1) \tag{2.29}$$

Next choose ω such that $\omega < \theta^{-m}$. Then relation (2.25) can be written as

$$\left\|\frac{dz}{dt} + A(t,x)z - f(\cdot,x)\right\|_{\infty,0} \le MQ^{-\mu}K^{\nu} \tag{2.30}$$

for some constant $M > 0$, since $\theta^{-m} = Q^{-\mu}$, by (2.27), (2.28). ∎

2.3 LEMMA The linearized equations (2.2) admit smooth approximate solutions of order (μ,ν,σ) with $\sigma = \nu$ in the sense of Definition 2.1.

Proof: The proof results from (2.30), (2.29), and (2.22). ∎

Thus we have shown that if $\bar{z}$ is a solution of the equation (2.16) that satisfies (2.17), then $z = S_\theta \bar{z}$ is an approximate solution of order (μ,ν,σ) with $\sigma = \nu$ of the linearized equation (2.2), that is, there exist a constant M and a function y such that

$$\frac{dz}{dt} + A(t,x)z - f(t,x) + y = 0 \tag{2.31}$$

$0 \le t \le b$, $z(0) = 0$, with y satisfying (2.4), where $\sigma = \nu$ and z satisfies (2.3), provided that $x \in G$ with $\|x\|_{\infty,p} < K$ $(K > 1)$, and η in (2.16) is chosen so as to satisfy relation (2.29).

Now put $z = x + h$. Then it follows from (2.5) that h is a solution of the equation

$$\frac{dh}{dt} + A(t,x)h + Px + y = 0 \qquad 0 \le t \le b \qquad h(0) = 0 \tag{2.32}$$

and we get from (2.4), (2.3), and (2.19) with $g = Px + y$ that

$$\|h\|_{\infty,0} \le bC(\|Px\|_{\infty,0} + MQ^{-\mu}K^{\nu}) \tag{2.33}$$

$$\|h\|_{\infty,p} \le \|x\|_{\infty,p} + MQK^{\nu} \quad ∎ \tag{2.34}$$

2.4 LEMMA Let $x \in G$ with $\|x\|_{\infty,p} < K$, $K > 1$, and let $0 < \bar{q} < q < 1$ be arbitrary fixed numbers. If h is a solution of the equation (2.32), then there exists $0 < \varepsilon \le 1$ such that

$$\|P(x + \varepsilon h) - (1 - \varepsilon)Px\|_{\infty,0} \le \varepsilon q\|Px\|_{\infty,0} \tag{2.35}$$

where $x + \varepsilon h \subset V_0$, and

$$\|h\|_{\infty,0} \le bC(1 + \bar{q})\|Px\|_{\infty,0} \tag{2.36}$$

Proof: We have the following identity

$$\begin{aligned} P(x + \varepsilon h) - (1 - \varepsilon)Px = {} & [A(\cdot, x + \varepsilon h) - A(\cdot, x)](x + \varepsilon h) \\ & - [f(\cdot, x + \varepsilon h) - f(\cdot, x)] \\ & + \varepsilon[dh/dt + A(t,x)h + Px + y] - \varepsilon y \end{aligned}$$

To obtain an estimate for the first two terms of the above identity we use relations (2.8) and (2.11) with small ε. The third term is equal to zero by (2.32). Hence, we obtain by (2.33) and (2.4) with $\sigma = \nu$,

$$\begin{aligned} \|P(x + \varepsilon h) - (1 - \varepsilon)Px\|_{\infty,0} & \le \varepsilon C(r + 1)\|h\|_{\infty,0} + \varepsilon MQ^{-\mu}K^{\nu} \\ & \le \varepsilon C(r + 1)bC\|Px\|_{\infty,0} + 2\varepsilon MQ^{-\mu}K^{\nu} \\ & \le \varepsilon(q - \bar{q})\|Px\|_{\infty,0} + 2\varepsilon MQ^{-\mu}K^{\nu} \\ & \le \varepsilon q\|Px\|_{\infty,0} \end{aligned}$$

provided b is so small that

$$C(r + 1)bC < q - \bar{q} < 1 \tag{2.37}$$

and the following relation holds, which also yields (2.36):

$$2MQ^{-\mu}K^{\nu} \le \bar{q}\|Px\|_{\infty,0} \quad \blacksquare \tag{2.38}$$

2.5 LEMMA Suppose that $x_n \in G$ and

$$\|x_n\|_{\infty,p} < A\exp(\alpha(1 - q)t_n) = K_n \tag{2.39}$$

where $t_n > 0$.

$$\mu(1 - \nu) - \nu > 0 \tag{2.40}$$

where $1 < \alpha, q, A$ are chosen so that

$$\alpha(1 - q) - 1 = \delta > 0 \quad \text{and} \quad \alpha > [\mu(1 - \nu) - \nu]^{-1} \tag{2.41}$$

and

$$M(2M)^{1/\mu}(\bar{q}p_0)^{-1/\mu} < A^{1-\nu-\nu/\mu}[\alpha(1 - q) - 1] \tag{2.42}$$

Then there exists a number $Q = Q_n$ such that

$$2MQ_n^{-\mu}K_n^{\nu} < \bar{q}p_0 \exp(-(1 - q)t_n) \tag{2.43}$$

and

$$\|x_n + \varepsilon h_n\|_{\infty,p} < A \exp(\alpha(1 - q)(t_n + \varepsilon)) \tag{2.44}$$

for all $0 < \varepsilon \leq 1$, where h_n is a solution of the equation (2.32) with $x = x_n$, $y = y_n$, which satisfies (2.34) and p_0 is as in (2.14).

Proof: We have by (2.34)

$$\begin{aligned}\|x_n + \varepsilon h_n\|_{\infty,p} &\leq \|x_n\|_{\infty,p} + \varepsilon\|h_n\|_{\infty,p} \\ &\leq (1 + \varepsilon)\|x_n\|_{\infty,p} + \varepsilon MQ_nK_n^{\nu}\end{aligned}$$

Hence, relation (2.44) is satisfied if

$$(1 + \varepsilon)A \exp(\alpha(1 - q)t_n) + \varepsilon MQK^{\nu} < A \exp(\alpha(1 - q)(t_n + \varepsilon)) \tag{2.45}$$

But it is easily seen that

$$\delta = \alpha(1 - q) - 1 = \min_{0\leq\varepsilon\leq 1} [\exp(\alpha(1 - q)\varepsilon) - (1 + \varepsilon)]/\varepsilon \tag{2.46}$$

Hence, by (2.46), $Q = Q_n$ is subject to

$$Q < (\alpha(1 - q) - 1)A^{1-\nu}M^{-1} \exp(\alpha(1 - \nu)(1 - q)t_n) \tag{2.47}$$

On the other hand, (2.43) yields

$$(2M)^{1/\mu}(\bar{q}p_0)^{-1/\mu}A^{\nu/\mu} \exp((\nu\alpha + 1)(1 - q)t_n/\mu) < Q \tag{2.48}$$

Hence, it follows that relations (2.47) and (2.48) are satisfied if conditions (2.40) to (2.42) hold. But this holds true provided that relation (2.29) is satisfied which implies that

$$Q^{\mu k} < \eta^{-\bar{k}} < K^{\nu}$$

Hence, $Q < K^{\nu/\mu\bar{k}}$, and we get

$$Q < A^{\nu/\mu\bar{k}} \exp(\alpha(1 - q)t_n\nu/\mu\bar{k}) \tag{2.49}$$

for $Q = Q_n$, $K = K_n$, by (2.39). Thus both relations (2.47) and (2.49) are satisfied if we assume that

$$1 - \nu < \frac{\nu}{\mu\bar{k}} \qquad \text{with} \qquad 0 < \nu < 1$$

or

$$\bar{k} < \frac{\nu}{(1 - \nu)\mu} \tag{2.50}$$

and A is so large that

$$(\alpha(1 - q) - 1)A^{1-\nu}M^{-1} < A^{\nu/\mu\bar{k}}$$

or, by (2.50),

$$(\alpha(1 - q) - 1)M^{-1} < A^{\beta} \qquad \beta = (\mu k)^{-1}\nu - (1 - \nu) > 0 \tag{2.51}$$

Let us observe that for relation (2.42) to be satisfied it is necessary that $1 - \nu - \nu/\mu > 0$ or $\mu > \nu/(1 - \nu)$. Hence, by (2.50),

$$\bar{k} < \frac{\nu}{\nu} = 1 \tag{2.52}$$

Now let $0 < \bar{k} < 1$ be arbitrary. Then $0 < \nu < 1$ can be chosen so as to satisfy (2.50). For the inequality (2.50) can be written as

$$\frac{\bar{k}\mu}{1 + \bar{k}\mu} < \nu < \frac{\mu}{1 + \mu} \tag{2.53}$$

by (2.40). ■

We can now construct the following method of contractor directions. Put $x_0(t) \equiv x_0 = 0$ and $t_0 = 0$. Then

$$x_{n+1} = x_n + \varepsilon_n h_n \qquad t_{n+1} = t_n + \varepsilon_n \tag{2.54}$$

with induction assumptions

$$\|x_n\|_{\infty,p} < A \exp(\alpha(1 - q)t_n) = K_n \tag{2.55}$$

and

$$\|Px_n\|_{\infty,0} \le p_0 \exp(-(1 - q)t_n) \tag{2.56}$$

where $p_0 = \|Px_0\|_{\infty,0} = \|f(\cdot,0)\|_{\infty,0}$ as in (2.14), h_n satisfies (2.32) with $x = x_n \in G$, $y = y_n$, $1 < \alpha, A, q$ being subject to (2.40), (2.41), (2.42), and (2.51) with $0 < \bar{k} < 1$ as in (2.17), and $0 < \nu < 1$ is chosen so as to satisfy (2.53). The verification of the induction assumptions (2.55) and (2.56) follows from Lemmas 2.4 and 2.5 (see Section 1). Since $z_n = x_n + h_n$, it is easily seen that (2.54) can be written as

$$x_{n+1} = (1 - \varepsilon_n)x_n + \varepsilon_n z_n \tag{2.54a}$$

which justifies the term "convex approximate linearization." The following estimates are valid for $\{x_n\}$.

$$\|h_n\|_{\infty,0} \le 2bCp_0 \exp(-(1 - q)t_n) \tag{2.57}$$

and with δ as in (2.46),

$$\|h_n\|_{\infty,p} \le (1 + \delta)A \exp(\alpha(1 - q)t_n) \tag{2.58}$$

In fact, relation (2.57) results from (2.36) with $x = x_n$, $h = h_n$, and (2.56). Relation (2.58) results from (2.34) with $x = x_n$, $h = h_n$, and (2.55) and (2.47).

In order to choose X_s put $\bar{s} = s/p$, and use (2.57), (2.58), and (A_0) to get $\|h_n\|_{\infty,s} \le C\|h_n\|_{\infty,0}^{1-\bar{s}}\|h_n\|_{\infty,p}^{\bar{s}}$, and consequently

$$\|h_n\|_{\infty,s} \le CNb^{1-\bar{s}} \exp(-\lambda(1 - q)t_n) \tag{2.59}$$

where

$$\begin{aligned} N &= (2Cp_0)^{1-\bar{s}}[(1 + \delta)A]^{\bar{s}} \\ &= (2Cp_0)^{1-\bar{s}}[\alpha(1 - q)A]^{\bar{s}} \end{aligned}$$

provided

$$\lambda = 1 - \bar{s}(1 + \alpha) > 0 \qquad \text{or} \qquad \alpha < \frac{1 - \bar{s}}{\bar{s}} \tag{2.60}$$

2.6 LEMMA The following estimate results from (2.59).

$$\sum_{n=0}^{\infty} \varepsilon_n\|h_n\|_{\infty,s} \le CNb^{1-\bar{s}}[(1 - q)\lambda]^{-1} \exp(\lambda(1 - q)) \tag{2.61}$$

Proof: For the proof see Section 1. ∎

Suppose that

$$CNb^{1-\bar{s}}[(1 - q)\lambda]^{-1} \exp(\lambda(1 - q)) < r \tag{2.62}$$

with N as in (2.59). Then there exists x such that

$$\|x_n - x\|_{\infty,s} \to 0 \quad \text{and} \quad \|x_n - x_0\|_{\infty,s} < r \tag{2.63}$$

Moreover, $\{Px_n\}$ is a Cauchy sequence and

$$\|Px_n - Px\|_{\infty,0} \to 0 \quad \text{as } n \to \infty \tag{2.64}$$

For the proof see Section 1.

2.7 LEMMA Suppose that condition (2.62) holds and $\{x_n\} \subset G$ is an arbitrary sequence which satisfies relations (2.63), (2.64), (2.55), and (2.56), where $t_n \to t < \infty$ as $n \to \infty$. Then there exist $0 < \varepsilon \le 1$ and $\bar{x}$ such that

$$\|P\bar{x} - (1 - \varepsilon)Px\|_{\infty,0} \le \varepsilon q' p_0 \exp(-(1 - q)t) \tag{2.65}$$

where $\bar{x} \in G$, $\|\bar{x} - x_0\|_{\infty,s} < r$; $\bar{q} < q' < q < 1$, $C < C_1 < \bar{C}$, $\bar{s} = s/p$, and

$$\|\bar{x} - x\|_{\infty,s} \le \varepsilon C_1 N_1 b^{1-\bar{s}} \exp(-\lambda(1 - q)t) \tag{2.66}$$

with N, s, and λ as in (2.59), (2.60), and N_1 is N with C replaced by C_1.

Proof: For the proof see Lemma 1.4. ∎

2.1 THEOREM In addition to the assumptions (A_0) to (A_5), suppose that ν is chosen so as to satisfy (2.53), and α > 1, s satisfy

$$[\mu(1 - \nu) - \nu]^{-1} < \alpha < \frac{p - s}{s}$$

$0 < \bar{q} < q < (\alpha - 1)/\alpha$, and A is chosen so as to satisfy (2.42) and (2.51), and $0 < b' \le b$ with $b\bar{C} < 1$ satisfies (2.37); also $C < \bar{C}$, C being as in (2.19), (A_0), and

$$\bar{C}N_1(b')^{1-\bar{s}}[\lambda(1 - q)]^{-1} \exp(\lambda(1 - q)) < r \tag{2.67}$$

where $\bar{s} = s/p$. Then the equation (2.1) with b replaced by b' has a solution x and the sequence $\{x_n\}$ determined by (2.54) converges to x, i.e.,

$$\|x_n - x\|_{\infty,s} \to 0 \quad \text{as } n \to \infty$$

$$\|x_n - x_0\|_{\infty,s} < r \quad \text{and} \quad \|x - x_0\|_{\infty,s} < r$$

where $x_0 = 0$.

Proof: The proof is exactly the same as in Theorem 1.1. ∎

2.1 REMARK Note that the completeness of X_p is not required. Hence, assuming that X_p is dense in X_{m_2}, one can take a vector space $V_p \subset X_p$ which is dense in X_p and consists of "smooth" vectors or functions, and replace V_0 by a new $V_0 = W_0 \cap V_p$. Then one assumes that g_ω in (2.12) is such that $g_\omega \in C(0,b;V_p(\|\cdot\|_p))$. Moreover, G can be made up of "smooth" vectors, by virtue of (2.54), if one assumes that $\bar{z}$ which is a solution of the equation (2.16) is "smooth" and that so is $z = S_\theta\bar{z}$, i.e., S_θ transforms V_p into V_p.

2.2 REMARK Suppose that there exists a sequence $\{A_m(t,x)\}$ of linear continuous mappings for which the following holds uniformly in G.

$$\|A_m(\cdot,x) - A(\cdot,x)\|_{\infty,X_s \to X_0} \to 0 \quad \text{as } m \to \infty \tag{i}$$

Then in (A_4), A(t,x) can be replaced by $A_m(t,x)$ with m sufficiently large. But then one has to assume that (2.16) and (2.17) hold with A(t,x) replaced by $A_m(t,x)$, uniformly in m for almost all m. The same applies to (A_5) with C independent of m. Then also in (2.9) and (2.10), A(t,x) should be replaced by $A_m(t,x)$, where C remains the same constant for almost all m. Then in the proof of Lemma 2.2 we get one more term, which is $\|[A(\cdot,x) - A_m(\cdot,x)]\bar{z}\|_{\infty,0} < \omega$, by (i), and we choose $2\omega < Q^{-\mu}$. The identity in Lemma 1.4 needs to be replaced by

$$\begin{aligned} P(x + \varepsilon h) - (1 - \varepsilon)Px = {} & [A(\cdot, x + \varepsilon h) - A(\cdot,x)](x + \varepsilon h) \\ & - [f(\cdot, x + \varepsilon h) - f(\cdot,x)] \\ & + \varepsilon[dh/dt + A_m(t,x)h + Px + y] - \varepsilon y \\ & + \varepsilon[A(t,x) - A_m(t,x)]h \end{aligned}$$

with m sufficiently large so as to get from (i) that

$$\|[A(\cdot,x) - A_m(\cdot,x)]h\|_{\infty,0} \le q'\|Px\|_{\infty,0}$$

with $q - \bar{q} - q' < 1$ and $C(r + 1)bC + q' < q - q'$, by (2.37). We use here the estimate (2.59). A similar change is needed in the proof of Lemma 2.7, where an identity like the above is employed, in which $A(t,x_n)h_n$ should be replaced by $A_m(t,x_n)h_n$ with sufficiently large m.

3. ELLIPTIC REGULARIZATION WITHOUT SMOOTHING OPERATORS

By imposing stronger conditions on the regularizing linear operator L, an existence theorem can be proved without making use of smoothing operators. Let $Z \subset E \subset Y \subset X$ be Banach spaces with norms $\|\cdot\|_X \le \|\cdot\|_Y \le \|\cdot\|_E \le \|\cdot\|_Z$ and let $W_0 \subset Y$ be an open ball with center $x_0 \in Z$ and radius $r > 0$. Put $V_0 = W_0 \cap Z$ and let V_1 be the $\|\cdot\|_Y$-closure of V_0. Using the same notation as in Section 1, let G be a set of functions $x \in C(0,b;V_0(\|\cdot\|_Z)) \cap C^1(0,b;X)$ with $x(0) = x_0$, $\|x - x_0\|_{\infty,Y} < r$, and $\|x\|_{\infty,Z} < \infty$.

Corresponding to each given $(t,x) \in [0,b] \times V_1$, let the mapping $A(t,x)$: $Y \to X$ be linear and continuous. Consider the Cauchy problem for the quasilinear equation

$$Px(t) \equiv \frac{dx}{dt} + A(t,x)x - f(t,x) = 0 \qquad 0 \le t \le b \qquad x(0) = x_0 \tag{3.1}$$

We make the following assumptions.

(A_0) There exist a number $0 < \bar{s} < 1$ and a constant $C > 0$ such that

$$\|x\|_Y \le C\|x\|_X^{1-\bar{s}}\|x\|_Z^{\bar{s}}$$

(A_1) $A(\cdot,x)$ and $f(\cdot,x)$ are uniformly continuous in the following sense. $\{x_n\} \subset G$ and $\|x_n - x\|_{\infty,Y} \to 0$ imply

$$\|A(\cdot,x_n) - A(\cdot,x)\|_{\infty,Y\to X} \to 0 \tag{3.2}$$

$$\|f(\cdot,x_n) - f(\cdot,x)\|_{\infty,X_0} \to 0 \tag{3.3}$$

as $n \to \infty$.

(A_2) There exists a constant $C > 0$ such that

$$\|[A(t, v + \varepsilon h) - A(t,v)]u\|_X \le C\varepsilon\|h\|_X\|u\|_Y \tag{3.4}$$

where $v + \varepsilon h \in V_0$; $(t,v) \in [0,b] \times V_0$, and

$$\|f(t, v + \varepsilon h) - f(t,v)\|_X \le C\varepsilon\|h\|_X \tag{3.5}$$

for all $(t,v) \in [0,b] \times V_1$, and $v + \varepsilon h \in V_1$. We also assume that $f(\cdot,x)$ can be approximated as follows. If $x \in G$, then for arbitrary $\omega > 0$, there exists a function g_ω such that

$$g_\omega \in C(0,b;Z(\|\cdot\|_Z))$$

and

$$\|f(\cdot,x) - g_\omega\|_{\infty,E} \le \omega \tag{3.6}$$

In addition, there exist d_0 and p_0 such that

$$\|f(\cdot,x)\|_{\infty,E} \le d_0 \qquad x \in G \tag{3.7}$$

and

$$\|A(\cdot,x_0)x_0 - f(\cdot,x_0)\|_{\infty,E} \le p_0 \tag{3.8}$$

(A_3) There exists a linear (regularizing) operator $L = L(\eta)$ such that

$$\|Lx\|_X \le C\|x\|_Z \tag{3.9}$$

for some constant $C > 0$.

There exists $k > 0$ such that the linearized equation

$$\frac{dz}{dt} + A(t,x)z + \eta Lz - g_\omega = 0 \tag{3.10}$$

$0 \le t \le b$, $z(0) = x_0$, with small $\eta > 0$ and $x \in G$, has a solution z such that

$$\|z\|_{\infty,Z} \le C\eta^{-k}\|g_\omega\|_{\infty,E} \tag{3.11}$$

for some $C > 0$.

(A_4) For $x \in G$ and g with $\|g\|_{\infty,X} < \infty$, if h is a solution of the equation

$$\frac{dh}{dt} + A(t,x)h + g = 0 \qquad 0 \le t \le b \qquad h(0) = 0 \tag{3.12}$$

then

$$\|h\|_{\infty,X} \le bC\|g\|_{\infty,X} \tag{3.13}$$

for some constant $C > 0$.

Consider the linearized equation

$$\frac{dz}{dt} + A(t,x)z - f(t,x) = 0 \qquad 0 \le t \le b \qquad z(0) = x_0 \tag{3.14}$$

and let z be a solution of the equation (3.10) satisfying (3.11).

3.1 LEMMA The following estimates hold for z.

$$\left\|\frac{dz}{dt} + A(\cdot,x)z - f(\cdot,x)\right\|_{\infty,X} \le (Cd_0 + 1)\eta^{-(k-1)} \tag{3.15}$$

and

$$\|z\|_{\infty,Z} \le C(d_0 + 1)\eta^{-k} \tag{3.16}$$

Proof: We have, for $x \in G$ and $\omega \le 1$,

$$\begin{aligned}\left\|\frac{dz}{dt} + A(\cdot,x)z - f(\cdot,x)\right\|_{\infty,X} &\le \left\|\frac{dz}{dt} + A(\cdot,x)z + \eta Lz - g_\omega\right\|_{\infty,X} \\ &\quad + \eta\|Lz\|_{\infty,X} + \|f(\cdot,x) - g_\omega\|_{\infty,X} \\ &\le \eta\|z\|_{\infty,Z} + \omega \le C\eta^{-(k-1)}\|g_\omega\|_{\infty,E} + \omega \\ &\le Cd_0\eta^{-(k-1)} + \omega \le (Cd_0 + 1)\eta^{-(k-1)}\end{aligned}$$

Relation (3.16) follows from (3.6), (3.7), (3.11), where $0 < \eta < 1$. ■

3.2 LEMMA Let $0 < k < 1$ be given, and let μ,ν be such that

$$\frac{1}{k} - 1 < \mu \quad \text{and} \quad 1 - [k(\mu + 1) - 1]^{-1} < \nu < \frac{\mu}{\mu + 1} \tag{3.17}$$

Then for arbitrary $K > 1$, one can find $0 < \eta < 1$ and $Q > 1$ such that

$$\eta^{-k} < QK^{\nu} \tag{3.18}$$

$$\eta^{-(k-1)} < Q^{-\mu}K^{\nu} \tag{3.19}$$

Proof: One can find a number $0 < \eta < 1$ such that

$$Q^{k(\mu+1)} < \eta^{-k} < QK^{\nu}$$

if $1 < Q < K^{\nu\beta}$ with $\beta = [k(\mu + 1) - 1]^{-1}$, which implies (3.18), by the first inequality (3.17). Hence,

$$\eta^{-(k-1)} = \eta\eta^{-k} < Q^{-(\mu+1)} \cdot QK^{\nu}$$

yielding (3.19). ∎

3.3 LEMMA The linearized equations (3.14) admit approximate solutions of order (μ,ν,σ) with $\sigma = \nu$ in the sense of Definition 1.1, where μ and ν satisfy (3.17), provided that relations (3.18) and (3.19) hold.

Proof: The proof follows immediately from the estimates (3.15) and (3.16) of Lemma 2.1 and from (3.18) and (3.19). One can put $M = C(d_0 + 1)$. ∎

It results from Lemma 3.3 that there exist functions z and y such that z is a solution of the equation

$$\frac{dz}{dt} + A(t,x)z - f(t,x) + y = 0 \tag{3.20}$$

$0 \le t \le b$, $z(0) = x_0$, and

$$\|z\|_{\infty,Z} \le MQK^{\nu} \tag{3.21}$$

$$\|y\|_{\infty,X} \le MQ^{-\mu}K^{\nu} \tag{3.22}$$

where $x \in G$ with $\|x\|_{\infty,Z} < K$.

Now put $z = x + h$. It is easily seen that h is a solution of the equation

$$\frac{dh}{dt} + A(t,x)h + Px + y = 0 \qquad 0 \le t \le b \qquad h(0) = 0 \tag{3.23}$$

In the same way as in Section 1 one can construct the following method of contractor directions.

Put $x_0(t) \equiv x_0$ and $t_0 = 0$. Then

$$x_{n+1} = x_n + \varepsilon_n h_n \qquad t_{n+1} = t_n + \varepsilon_n \tag{3.24}$$

with induction assumptions

$$\|x_n\|_{\infty,Z} < A \exp(\alpha(1 - q)t_n) = K_n \tag{3.25}$$

and

$$\|Px_n\|_{\infty,X} \le p_0 \exp(-(1 - q)t_n) \tag{3.26}$$

where $\|Px_0\|_{\infty,X} = \|A(\cdot,x_0)x_0 - f(\cdot,x_0)\|_{\infty,X} \le p_0$, h_n is a solution of (3.23) with $x = x_n$.

The choice of α,A is exactly the same as in Section 2 except for (2.49), which by (3.19) should be replaced by

$$Q < K^{\nu\beta} \qquad \text{with} \qquad \beta = [k(\mu + 1) - 1]^{-1}$$

that is,

$$Q < A^{\nu\beta} \exp(\nu\beta\alpha(1 - q)t_n) \tag{3.27}$$

which is satisfied along with (2.47) if $1 - \nu < \nu\beta$. But this inequality is true if the second inequality (3.17) is satisfied. By the same argument, $A^{1-\nu} < A^{\nu\beta}$. Thus condition (2.50) is not needed here.

The choice of $\{\varepsilon_n\}$ is exactly the same as in Section 1. Note that X_0, X_s, X_p should be replaced by X, Y, Z, respectively. With this in mind the reasoning presented in Section 1 can be repeated. We are now in a position to prove the following.

3.1 THEOREM In addition to assumptions (A_0) to (A_4), suppose that μ and ν are chosen so as to satisfy (3.17). Also assume that (2.67) holds. Then equation (3.1) with b replaced by b' has a solution x which is a limit of $\{x_n\}$ determined by (3.23), i.e., $\|x_n - x\|_{\infty,Y} \to 0$ as $n \to \infty$, and

$$\|x_n - x_0\|_{\infty,Y} < r \qquad \|x - x_0\|_{\infty,Y} < r$$

Proof: The proof is exactly the same as in Theorem 1.1. ∎

3.1 REMARK Remarks 2.1 and 2.2 also apply to Theorem 3.1.

An Application of C_0-Semigroups

Sufficient conditions for assumption (A_5) in Section 1 to be satisfied can be given by making use of the theory of C_0-semigroups.

Denote by $G(X_0)$ the set of all negative infinitesimal generators of C_0-semigroups $\{U(t)\}$ on X_0. If $-A \in G(X_0)$ then $\{U(t)\} = \{e^{-tA}\}$, $0 \le t < \infty$, is the semigroup generated by $-A$. For $x \in G$ put $A(t) = A(t,x(t))$. We assume that A is a function from $[0,b] \times V_0$ into $G(X_0)$ and A is stable in X_0 (see Kato, 1970; 1973; 1975; Yosida, (1968; Tanabe, 1979). Moreover, operators $U(t,s;x)$ generated by $A(t,x(t))$ exist and satisfy the following condition:

$$\|U(t,s;x)\|_{X_0} \le M_1$$

for all $x \in G$ and some constants $M_1 > 0$. An alternative condition is that the approximate evolution operators $U_m(t,s;x)$ generated by the step functions $A_m(t,x(t))$ are uniformly convergent to $A(t,x(t))$ and satisfy the following condition

$$\|U_m(t,s;x)\|_{X_0} \le M_1$$

for all $x \in G$ and some constant $M_1 > 0$.

What is said above also applies to Section 2, where X_0, X_s are replaced by X,Y, respectively.

3.2 REMARK For a better choice of μ, instead of (2.28) one can use Lemma 1.6 or 1.6a, Chapter 2, and $i = m_2$ in (2.17).

11

Quasilinear Evolution Equations

As a result of our previous investigations a new approach is presented for solving abstract quasilinear equations in nonreflexive Banach spaces. This approach is also based on the theory of contractor directions (Altman, 1981; 1983a, b) and is not based directly on the C_0-semigroups theory. The general ideas developed in Altman (1984a) are exploited here and lead to a rather surprising result. It turns out that by imposing an extra Lipschitz condition on the family $\{A(t,x)\}$ which appears in the evolution equations under consideration, one can, at least theoretically, eliminate the restriction that requires the existence of a family of isomorphisms $\{S(t)\}$ (see Kato, 1975). Kato's theory is based on C_0-semigroups and applies to reflexive Banach spaces only.

1. CONVEX LINEARIZATION AND A METHOD OF CONTRACTOR DIRECTIONS

Let X, Y, Z be Banach spaces and denote by $\|\cdot\|_X$, $\|\cdot\|_Y$, $\|\cdot\|_Z$ the norms in X, Y, Z, respectively. We assume that the embedding $Z \subset Y \subset X$ is continuous and Z is dense in X. Let W_0 be an open ball in Y with center $x_0 \in Z$ and radius $r > 0$, and denote by W its closure in Y, and let Z_0 be an open ball in Z with the same center x_0 and radius

$R > 0$. Given $0 < b$, denote by $C(0,b;X)$, $C(0,b;Y)$, $C(0,b;Z)$ the spaces of all continuous functions $x = x(t)$, $y = y(t)$, $z = z(t)$, defined on the interval $[0,b]$ with values in X, Y, and Z and norms

$$\|x\|_{C(X)} = \sup_t [\|x(t)\|_X: \quad 0 \le t \le b]$$

$$\|y\|_C = \sup_t [\|y(t)\|_Y: \quad 0 \le t \le b]$$

$$\|z\|_D = \sup_t [\|z(t)\|_Z: \quad 0 \le t \le b]$$

respectively. Denote by $C^1(0,b;X)$ the vector space of all continuously differentiable functions from $[a,b]$ into X. In what follows Z may be incomplete.

We assume that there exists a vector space $V \subset Z$ which is dense in $Z(\|\cdot\|_Z)$ and has the following property. Corresponding to each $t \in [0,b]$ and x contained in the norm $\|\cdot\|_Y$-closure of $V_0 = V \cap Z_0 \cap W_0$, the mapping $A(t,x)$: $Y \to X$ is linear and continuous. We may assume that $\|\cdot\|_Z \ge \|\cdot\|_Y$ and $\|x_0\|_Z < R$.

Consider the Cauchy problem for the quasilinear equation

$$Px(t) \equiv \frac{dx}{dt} + A(t,x)x - f(t,x) = 0 \qquad (1.1)$$
$$0 \le t \le b \qquad x(0) = x_0$$

We make the following assumptions.

(A_1) $f(t,v) \in Z$ for all $(t,v) \in [0,b] \times V_0$, and $f(t,x) \in Y$ for all $t \in [0,b]$ and x contained in the $\|\cdot\|_Y$-closure of V_0. If $(t,x) \in [0,b] \times V_0$, then $A(t,x)v \in Z$ for all $v \in V$. We assume that $x_0 = 0$ and p_0 is such that

$$\|A(t,x_0)x_0 - f(t,x_0)\|_Y \le p_0 \qquad \text{for all } 0 \le t \le b \qquad (1.2)$$

(A_2) Let $x \in C(0,b;V_0(\|\cdot\|_Z)) \cap C^1(0,b;X)$ with $x(0) = 0$ and $y \in L^1(0,b;V(\|\cdot\|_Z)) \cap C(0,b;X)$.

Then the equation

$$\frac{dz}{dt} + A(t,x)z - y(t) = 0 \qquad 0 \le t \le b \qquad z(0) = 0 \qquad (1.3)$$

has a solution $z \in C(0,b;V(\|\cdot\|_Z)) \cap C^1(0,b;X)$ such that

$$\|z\|_D \le bC\|y\|_D \tag{1.4}$$

where C is a global constant; and if $z = x + h$, then h is obviously a solution of the equation

$$\frac{dh}{dt} + A(t,x)h + \left[\frac{dx}{dt} + A(t,x)x - y\right] = 0 \tag{1.5}$$

$0 \le t \le b$, $h(0) = 0$, and we assume

$$\|h\|_C \le bC\left\|\frac{dx}{dt} + A(\cdot,x)x - y\right\|_C \tag{1.6}$$

whenever $\|dx/dt + A(\cdot,x)x - y\|_C \le 2p_0$.

(A_3) There exists a constant K_2 such that

$$\|[A(t,u) - A(t,v)]y\|_X \le K_2\|u - v\|_Y\|y\|_Y \tag{1.7}$$

for all $t \in [0,b]$ and u,v contained in the $\|\cdot\|_Y$-closure of V_0, and $y \in Y$.

(A_4) There exists a constant K_3 such that

$$\|f(t,u) - f(t,v)\|_Y \le K_3\|u - v\|_Y \tag{1.8}$$

for all $t \in [0,b]$ and u,v contained in the $\|\cdot\|_Y$-closure of V_0.

We also assume that $f(\cdot,x)$ can be approximated as follows. Let

$$x \in C(0,b;V_0(\|\cdot\|_Z)) \cap C^1(0,b;X)$$

Then for arbitrary $\omega > 0$, there exists a function

$$y_\omega \in L^1(0,b;V(\|\cdot\|_Z)) \cap C(0,b;X)$$

such that

$$\|f(\cdot,x) - y_\omega\|_D < \omega \tag{1.9}$$

and we also assume that d_0 is such that

$$\|f(\cdot,v)\|_D \le d_0 \qquad \text{for all } v \in V_0 \tag{1.10}$$

Finally, we assume the following extra Lipschitz condition.

(A_5) There exists a constant K_1 such that

$$\|[A(t, v + \varepsilon h) - A(t,v)]u\|_Y \le \varepsilon K_1 \|h\|_Y \|u\|_Z \tag{1.11}$$

for all $t \in [0,b]$, $v \in V_0$, $h,u \in V$, and $0 < \varepsilon \le 1$ such that $v + \varepsilon h \in V_0$.

One can weaken this hypothesis by replacing v in (1.11) by

$$x \in C(0,b;V_0(\|\cdot\|_Z)) \cap C^1(0,b;X) \qquad x(0) = x_0$$

where $z = x + h$ is a solution of equation (1.3).

Let us mention that if (1.9) can be satisfied with $y_\omega \in C(0,b;V(\|\cdot\|_Z))$, then also in ($A_2$) we assume that $y \in C(0,b;V(\|\cdot\|_Z))$.

1.1 LEMMA If b is sufficiently small, then $\|z - x_0\|_D < R$ whenever z is a solution of equation (1.3), where $x(0) = x_0$ and $x \in C(0,b;V_0(\|\cdot\|_Z)) \cap C^1(0,b;X)$ and y satisfies relation (1.9).

Proof: We have, by virtue of (1.4),

$$\|z\|_D \le b\|y\|_D \le bC(\|f(\cdot,x) - y\|_D + \|f(\cdot,x)\|_D)$$
$$\le bC(d_0 + d_0) < R - \|x_0\|_Z$$

if b is sufficiently small and, say, $\omega < d_0$, by virtue of (1.10).

We assume that b is so small that the hypotheses of Lemma 1.1 are satisfied. ∎

1.2 LEMMA There exists a constant K_4 such that

$$\|[A(\cdot, x + \varepsilon h) - A(\cdot,x)](x + \varepsilon h)\|_C \le \varepsilon K_4\|h\|_C \tag{1.12}$$

where $x \in C(0,b;V_0(\|\cdot\|_Z)) \cap C^1(0,b;X)$, $x(0) = x_0$, and $z = x + h$ is a solution of equation (1.3) and $0 < \varepsilon \le 1$ is such that $x + \varepsilon h$ is contained in W_0.

Proof: Since $x + \varepsilon h = (1 - \varepsilon)x + \varepsilon z$, it follows from Lemma 1.1 and assumption (1.11) that relation (1.12) holds with $K_4 = K_1(\|x_0\|_Z + R)$. ∎

Now let $x \in C(0,b;V_0(\|\cdot\|_Z)) \cap C^1(0,b;X)$, $x(0) = 0$, with Px contained in Y and $\|Px\|_C < \infty$. Let $y \in L^1(0,b;V(\|\cdot\|_Z)) \cap C(0,b;X)$ satisfy (1.9) with small ω to be determined. Let $z = x + h$ be a solution of equation (1.3) with h satisfying equation (1.5). Then, we obtain, by virtue of (1.6),

$$\begin{aligned}\|h\|_C &\le bC\|dx/dt + A(\cdot,x)x - y\|_C \\ &\le bC(\|Px\|_C + \|y - f(\cdot,x)\|_C) \\ &\le bC_1\|Px\|_C\end{aligned}$$

if $C < C_1$ and ω is sufficiently small. Hence, we get

$$\|h\|_C \le bC_1\|Px\|_C \tag{1.13}$$

Now put $\bar{x} = x + \varepsilon h$, where $0 < \varepsilon \le 1$ is so small that $\bar{x}$ is contained in W_0. Thus by virtue of (1.13), it suffices to choose ε so as to satisfy

$$\varepsilon bC_1\|Px\|_C < r - \|x - x_0\|_C$$

since x is already contained in W_0 by assumption. Furthermore, let $0 < b' \le b$ and q be such that

$$\bar{q} = b'C_1(K_4 + K_3) < q < 1 \qquad \text{and} \qquad C < C_1 \tag{1.14}$$

1.3 LEMMA The following relations hold.

$$\|P(x + \varepsilon h) - (1 - \varepsilon)Px\|_C \le q\varepsilon\|Px\|_C \tag{1.15}$$

$$\|h\|_C \le b'C_1\|Px\|_C \tag{1.16}$$

and $\bar{x} = x + \varepsilon h$ satisfies the following relation:

$$\bar{x} \in C(0,b';V_0(\|\cdot\|_Z)) \cap C^1(0,b;X) \qquad \bar{x}(0) = x_0 \tag{1.17}$$

provided that $\omega \le (\bar{q} - q)\|Px\|_C$.

Proof: We have the following identity:

$$\begin{aligned}P(x + \varepsilon h) - Px + \varepsilon Px = {} & [A(\cdot, x + \varepsilon h) - A(\cdot,x)](x + \varepsilon h) \\ & - [f(\cdot, x + \varepsilon h) - f(\cdot,x)]\end{aligned}$$

$$+ \varepsilon[dh/dt + A(t,x)h + dx/dt$$
$$+ A(\cdot,x)x - y] + \varepsilon[y - f(\cdot,x)]$$

To obtain an estimate for the first term of the above identity we use relation (1.12) of Lemma 1.2 and (1.16). An estimate for the second term follows from relations (1.8) and (1.16). The third term is equal to zero. An estimate for the fourth term follows from (1.9) with $\omega \le (\bar{q} - q)\|Px\|_C$. In this way we obtain the following estimates by combining the first two terms of the above identity and using (1.14).

$$\|P(x + \varepsilon h) - Px + \varepsilon Px\|_C \le \varepsilon b' C_1(K_4 + K_3)\|Px\|_C + \varepsilon(q - \bar{q})\|Px\|_C$$
$$= \varepsilon\bar{q}\|Px\|_C + \varepsilon(q - \bar{q})\|Px\|_C$$

Relation (1.16) results from (1.13) with b replaced by b'. Furthermore, we have

$$\bar{x} = x + \varepsilon h = (1 - \varepsilon)x + \varepsilon z$$

Since $\bar{x}$ is contained in W_0, we conclude from the above that $\bar{x}$ is contained in Z_0 because so is z by assumption. It is now clear that relation (1.17) holds. ∎

1.4 LEMMA Let the sequence

$$\{x_n\} \subset C(0,b';V_0(\|\cdot\|_Z)) \cap C^1(0,b';X) \tag{1.17a}$$

be such that

$$\|x_n - x\|_C \to 0 \qquad \|Px_n - Px\|_C \to 0 \qquad \|x - x_0\|_C < r \tag{1.18}$$

Let q', C_1 be such that

$$\bar{q} < q' < q \quad \text{and} \quad b'C_1 < \bar{C}_1 < \bar{C} \tag{1.19}$$

[see (1.14)]. Then there exist $0 < \varepsilon \le 1$ and

$$\bar{x} \in C(0,b';V_0(\|\cdot\|_Z)) \cap C^1(0,b';X) \tag{1.20}$$

such that

$$\|P\bar{x} - (1 - \varepsilon)Px\|_C \le \varepsilon q'\|Px\|_C \tag{1.21}$$

$$\|\bar{x} - x\|_C \le \varepsilon\bar{C}_1\|Px\|_C \tag{1.22}$$

Proof: Choose $0 < \varepsilon \le 1$ such that

$$\varepsilon b'C_1\|Px\|_C < r - \|x - x_0\|_C \tag{1.23}$$

and n so large that

$$\|x_n - x\|_C < r - \|x - x_0\|_C - \varepsilon b'C_1\|Px\|_C \tag{1.24}$$

Having chosen $0 < \varepsilon \le 1$, we then move with x_n, Px_n closer to x, Px, respectively, so as to satisfy the following relations:

$$\|Px_n - Px\|_C \le \varepsilon\eta\bar{q}\|Px\|_C \qquad \eta = (q' - \bar{q})/3\bar{q} \tag{1.25}$$

where $\bar{q}$ is given by (1.14), and

$$\|x_n - x\|_C \le \varepsilon(\bar{C}_1 - b'C_1)\|Px\|_C \tag{1.26}$$

Let $y_n \in L^1(0,b';V(\|\cdot\|_Z)) \cap C(0,b';X))$ be such that

$$\|y_n - f(\cdot,x_n)\|_D \le \omega \le \eta\bar{q}\|Px\|_C \tag{1.27}$$

with η as in (1.25). Now let $z_n = x_n + h_n$ be a solution of the equation

$$\frac{dz_n}{dt} + A(t,x_n)z_n - y_n = 0 \tag{1.28}$$

$0 \le t \le b' \le b$, $z_n(0) = x_0$. Then we get

$$\frac{dh_n}{dt} + A(t,x_n)h_n + \frac{dx_n}{dt} + A(\cdot,x_n)x_n - y_n = 0 \tag{1.28}$$

$h_n(0) = 0$, and by virtue of (1.6), we obtain

$$\begin{aligned}\|h_n\|_C &\le b'C\|dx_n/dt + A(\cdot,x_n)x_n - y_n\|_C\\ &\le b'C(\|Px_n\|_C + \|y_n - f(\cdot,x_n)\|_C)\\ &\le b'C(\|Px\|_C + \|Px_n - Px\|_C + \|y_n - f(\cdot,x_n)\|_C)\\ &\le b'C_1\|Px\|_C\end{aligned}$$

Hence, we obtain for $C_1 > C$,

$$\|h_n\|_C \le b'C_1\|Px\|_C \tag{1.29}$$

if n is sufficiently large to satisfy relations (1.24) to (1.27). In order to prove relation (1.21) for $\bar{x} = x_n + \varepsilon h_n$, the following identity is used.

$$\begin{aligned}
P\bar{x} - (1 - \varepsilon)Px &= P(x_n + \varepsilon h_n) - (1 - \varepsilon)Px_n + (1 - \varepsilon)[Px_n - Px] \\
&= [A(\cdot, x_n + \varepsilon h_n) - A(\cdot, x_n)](x_n + \varepsilon h_n) \\
&\quad - [f(\cdot, x_n + \varepsilon h_n) - f(\cdot, x_n)] \\
&\quad + \varepsilon[dh_n/dt + A(t, x_n)h_n + dx_n/dt + A(\cdot, x_n)x_n - y_n] \\
&\quad + \varepsilon[y_n - f(\cdot, x_n)] + (1 - \varepsilon)[Px_n - Px]
\end{aligned}$$

To obtain an estimate for the first term of the above identity we use Lemma 1.2 with $x = x_n$ and $h = h_n$. An estimate for the second term follows from (1.8). The third term is equal to zero. An estimate for the fourth term is given by relation (1.27). Hence, we obtain the following estimate by combining the first two terms of the above identity:

$$\begin{aligned}
\|P(x_n + \varepsilon h_n) - (1 - \varepsilon)\|Px\|_C &\le \varepsilon b'C_1(K_4 + K_3)\|Px\|_C \\
&\quad + \varepsilon 2\eta\bar{q}\|Px\|_C + \varepsilon\eta\bar{q}\|Px\|_C \\
&= \varepsilon\bar{q}\|Px\|_C + 3\varepsilon\eta\bar{q}\|Px\|_C \\
&\le \varepsilon q'\|Px\|_C
\end{aligned}$$

by virtue of (1.25), (1.29), (1.14), and (1.19). Hence, we conclude that if $n = \bar{n}$ is so large that relations (1.24) to (1.27) are satisfied, then relation (1.21) holds. Relation (1.22) results from the following inequality:

$$\begin{aligned}
\|\bar{x} - x\|_C &\le \|x_n - x\|_C + \varepsilon\|h_n\|_C \\
&\le \varepsilon(\bar{C}_1 - b'C_1)\|Px\|_C + \varepsilon b'C_1\|Px\|_C \\
&= \varepsilon\bar{C}_1\|Px\|_C
\end{aligned}$$

by virtue of (1.26) and (1.29). Finally, since $\bar{x} = (1 - \varepsilon)x_n + \varepsilon z_n$, $n = \bar{n}$, relation (1.20) holds, by virtue of relation (1.17a) and Lemma 1.3 with $x = x_n$ and $h = h_n$, since $\bar{x}$ is in W_0, by (1.24) and (1.29). ■

1.5 LEMMA Let the sequence

$$\{x_n\} \subset C(0,b';V_0(\|\cdot\|_Z)) \cap C^1(0,b';X)$$

be such that

$$\|x_n - x\|_C \to 0 \quad \text{and} \quad \|Px_n - z\|_C \to 0 \quad \text{as } n \to \infty \tag{1.30}$$

Then $z = Px$.

Proof: It is easily seen from assumption (A_3) that there exists a constant k such that

$$\|A(t,x)y\|_X \le k\|y\|_Y \tag{1.31}$$

for all $t \in [0,b']$ and $y \in Y$. Since

$$\|A(t,x_n(t))x_n(t) - A(t,x(t))x(t)\|_X$$
$$\le \|[A(t,x_n(t)) - A(t,x(t))]x_n(t)\|_X$$
$$+ \|A(t,x(t))[x_n(t) - x(t))]\|_X$$

it results from (1.7) and (1.31) that

$$A(\cdot,x_n)x_n \to A(\cdot,x)x \quad \text{as } n \to \infty \tag{1.32}$$

where the convergence is in the norm $\|\cdot\|_X$ and uniform in $t \in [0,b']$. It follows from relation (1.8) that

$$f(\cdot,x_n) \to f(\cdot,x) \quad \text{as } n \to \infty \tag{1.33}$$

where the convergence is in the norm $\|\cdot\|_C$. Hence, we see, by virtue of (1.30) to (1.33), that the sequence $\{dx_n/dt\}$ is convergent, and consequently, by virtue of (1.30), we get

$$\frac{dx_n}{dt} \to \frac{dx}{dt} \quad \text{as } n \to \infty \tag{1.34}$$

where the convergence is in the norm $\|\cdot\|_X$ and uniform in $t \in [0,b']$. Finally, it is clear from relations (1.30) to (1.34) that $Px_n \to Px$ as $n \to \infty$. Hence, we obtain that $z = Px$. ∎

We are now in a position to prove the following.

1.1 THEOREM In addition to the hypotheses (A_1) to (A_5), suppose that the radius r of the open ball $W_0 \subset Y$ with center $x_0 \in V_0$ satisfies the following relation:

$$r > \bar{r} = (1 - q)^{-1}\bar{C}p_0 \exp(1 - q) \qquad bC < \bar{C}$$

where p_0, C, $0 < q < 1$ are determined by (1.2), (1.6), and (1.14), respectively. Then there exist $0 < b' \le b$ and a continuous function $x \in C(0,b';W(\|\cdot\|_Y)) \cap C^1(0,b';X)$ satisfying equation (1.1) for all $t \in [0,b']$.

Proof: We construct a sequence

$$\{x_n\} \subset C(0,b';V_0(\|\cdot\|_Z)) \cap C^1(0,b';X) \tag{1.35}$$

and a sequence $\{\varepsilon_n\}$ with $0 < \varepsilon_n \le 1$ such that

$$\|Px_{n+1} - (1 - \varepsilon_n)Px_n\|_C \le \varepsilon_n q\|Px_n\|_C \tag{1.36}$$

$$\|x_{n+1} - x_n\|_C \le \varepsilon_n\bar{C}\|Px_n\|_C \tag{1.37}$$

as follows. Put $x_0(t) \equiv x_0 \in V_0$, and suppose that $x_0, x_1, \ldots, x_n$ and $\varepsilon_0, \varepsilon_1, \ldots, \varepsilon_{n-1}$ are known. It follows from Lemma 1.3 with $x = x_n$ that there exist $0 < \bar{\varepsilon}_n \le 1$ and $\bar{x}_{n+1} = x_n + \bar{\varepsilon}_n h_n$ such that

$$\begin{aligned} &\bar{x}_{n+1} \in C(0,b';V_0(\|\cdot\|_Z)) \cap C^1(0,b;;X) \\ &\|P\bar{x}_{n+1} - (1 - \bar{\varepsilon}_n)Px_n\|_C \le \bar{\varepsilon}_n q\|Px_n\|_C \\ &\|\bar{x}_{n+1} - x_n\|_C \le \bar{\varepsilon}_n\bar{C}\|Px_n\|_C \end{aligned} \tag{1.38}$$

Then we choose ε_n to be close to the largest number for which x_{n+1} exists and satisfies relation (1.35) with x_n replaced by x_{n+1} as well as relations (1.36) and (1.37). Now it follows from Lemma 2.1,

Appendix 1, that $\|x_n - x_0\|_C < \bar{r} < r$ for all n, and there exist x and z such that

$$\|x_n - x\|_C \to 0 \quad \text{and} \quad \|Px_n - z\|_C \to 0 \quad \text{as } n \to \infty$$

By virtue of Lemma 1.5, we have that $z = Px$. It also follows from Lemma 2.1 that if $\{\varepsilon_n\}$ is not convergent to 0, then $Px = 0$. Therefore suppose $Px \neq 0$ and

$$\varepsilon_n \to 0 \quad \text{as } n \to \infty \tag{1.39}$$

It results from Lemma 1.4 that there exist $0 < \varepsilon \leq 1$ and $\bar{x}$ which satisfy relations (1.20) to (1.22). Since

$$\|x_n - x\|_C \to 0 \quad \text{and} \quad \|Px_n - Px\|_C \to 0 \quad \text{as } n \to \infty$$

and the sequence $\{\|Px_n\|_C\}$ is decreasing, it follows that

$$\|P\bar{x} - (1 - \varepsilon)Px_n\|_C \leq \varepsilon q\|Px_n\|$$

and

$$\|\bar{x} - x_n\|_C \leq \varepsilon\bar{C}\|Px_n\|_C$$

for all sufficiently large n. But the maximal choice of ε_n implies that $\varepsilon_n \geq \varepsilon/2$ for all sufficiently large n. But this is a contradiction to (1.39), and the proof of the theorem is completed. ∎

1.1 REMARK We can make the following extra assumption in (A_2) in order to guarantee that $dx/dt + A(\cdot,x)x - y \in L^1(0,b;V(\|\cdot\|_Y)) \cap C(0,b;X)$ whenever in the proofs we deal with equation (1.5).

(a) Assume that $x,z \in C(0,b;V_0(\|\cdot\|_Z)) \cap C^1(0,b;X)$ implies $A(\cdot,x)z \in L^1(0,b;V(\|\cdot\|_Y))$.

(b) We assume that for each $v \in V_0$, the function $t \to A(t,v)$ is continuous in the $B(Y,X)$ norm.

Then it follows from (1.7) that $A(\cdot,x)z \in C(0,b;X)$.

(c) Relation (1.6) is supposed to hold provided that $dx/dt \in L^1(0,b;V(\|\cdot\|_Y))$.

It follows from the above and from Lemma 1.1 that under the hypotheses of Lemma 1.3 if $dx/dt \in L^1(0,b;V(\|\cdot\|_Y))$, then the same condition is true for $d\bar{x}/dt$, that is, $d\bar{x}/dt \in L^1(0,b;V(\|\cdot\|_Y))$, where $\bar{x} = x + \varepsilon h$. In fact $\bar{x} = (1 - \varepsilon)x + \varepsilon z$, and we have by assumption that $y, A(\cdot,x)z \in L^1(0,b;V(\|\cdot\|_Y))$. Hence, $dz/dt \in L^1(0,b;V(\|\cdot\|_Y))$, by virtue of (1.3).

It should be mentioned that also in Lemma 1.4 the condition $dx_n/dt \in L^1(0,b;V(\|\cdot\|_Y))$ should be added to (1.7a) and $d\bar{x}/dt \in L^1(0,b;V(\|\cdot\|_Y))$ to (1.21) and (1.22). The same remark applies to the proof of Theorem 1.1; that is, the condition $dx_n/dt \in L^1(0,b;V(\|\cdot\|_Y))$ should be added to relation (1.35), and $d\bar{x}_{n+1}/dt \in L^1(0,b;V(\|\cdot\|_Y))$ to relation (1.38).

1.2 REMARK Since we deal with

$$\|dx/dt + A(t,x)x - y\|_Y \leq 2p_0$$

it suffices to assume in (a) that $A(\cdot,x)z$ is Y-measurable. Then we replace the corresponding relations of integrability by Y-measurability, e.g., $d\bar{x}/dt$ is Y-measurable.

Kato's (1970) theory of evolution operators for solving abstract linear evolution equations can be used to obtain sufficient conditions for assumptions (A_2) to be satisfied.

(i) In what follows Z is assumed to be complete and $x_0 = 0$. Suppose that the family of linear bounded operators $\{A(t,x)\} \subset B(Z,X)$ satisfies Kato's (1970) hypotheses so that the evolution operators $\{U(t,s;x)\}$ exist and the solution of equation (1.3) is given by the formula

$$z(t) = \int_0^t U(t,s;x)y(s)\, ds \tag{S}$$

In addition, we assume that

$$\|U(\cdot,\cdot;x)\|_{\infty,Z} \leq M$$

for some constant M and all $x \in z_0$, using Kato's (1975) notation.

Then using Kato's estimate we obtain

$$\|z\|_D = \|z\|_{\infty,Z} \leq M \int_0^b \|y(t)\|_Z \, dt \tag{*}$$

Hence, we obtain relation (1.4).

(ii) Suppose that we replace in (i) Z by Y and Z_0 by W_0 and we assume that similar assumptions are made with regard to the family $\{A(t,x)\} \subset B(Y,X)$.

(iii) In addition, we assume that if

$$x, z \in C(0,b;V_0(\|\cdot\|_Z)) \cap C^1(0,b;X)$$

then the function $A(\cdot,x)z \in L^1(0,b;V(\|\cdot\|_Y))$.

(iv) We also assume that for each $v \in V_0$, the function $t \to A(t,v)$ is continuous in the $B(Y,X)$ norm. This implies, by virtue of (1.7), that $A(\cdot,x)z \in C(0,b;X)$, and we can apply Kato's estimate to obtain relation (1.6).

Along with Kato's theory of evolution operators, Remark 1.1 has to be applied and Banach spaces $Z \subset Y$ should be chosen so as to satisfy (A_5). The remaining hypotheses are without change.

Under the above hypotheses the proof of Theorem 1.1 carries over with minor modifications indicated in Remark 1.1.

Remark 1.2 also applies here.

EXAMPLE The following example is given by Kato (1975).

Consider the quasilinear equation

$$u_t + a(u)u_x = b(u)u \qquad t > 0 \qquad -\infty < x < \infty$$

where $u = u(t,x)$ is real valued and $u_t = \partial u/\partial t$, $u_x = \partial u/\partial x$. The real-valued functions a and b are assumed to be sufficiently smooth. We choose

$$X = H^0(-\infty,\infty) \qquad Y = H^{s_1}(-\infty,\infty) \qquad Z = H^{s_2}(-\infty,\infty)$$

$s_2 > s_1$, where H^s denotes the Sobolev space of order s of the L^2-type, so that $H^0 = L^2$. The vector space $V \subset Z$ consists of sufficiently smooth functions of $H^{s_2}(-\infty,\infty)$.

We put $D = d/dx$, and

$$A(t,y) = A(y)D \qquad f(t,y) = f(y) = f(y) = yb(y)$$

Kato (1975) has shown that his hypotheses are satisfied, which imply the existence of the evolution operator $U(t,s;x)$ for solving equations (1.3) by the formula (S). This yields the estimate given by (*) to obtain (1.4). Kato also noticed that $Y = H^s(-\infty,\infty)$ may be chosen with any integer $s \geq 2$. Hence, Kato's hypotheses are also satisfied for $s_2 > s_1 \geq 2$, so that the estimate given by (*) can be applied to the solution of equation (1.5), yielding (1.6), because assumption (iii) can obviously be satisfied for $A(t,y) = A(y) = A(y)D$. It is easily seen that $s_2 > s_1$ can be chosen so as to satisfy the extra Lipschitz condition (1.11) in (A_5). The remaining assumptions can be verified without difficulty.

1.3 REMARK Theorem 1.1 yields only existence of a solution of the quasilinear evolution equation. However, it is easily seen that a uniqueness theorem can be obtained if we assume the following stronger Lipschitz conditions.

$$\|[A(t,u) - A(t,v)]y\|_X \leq K_1\|u - v\|_X\|y\|_Y$$

and

$$\|f(t,u) - f(t,v)\|_X \leq K_2\|u - v\|_X$$

for some $K_1, K_2 > 0$ and all u,v in the $\|\cdot\|_Y$-closure of V_0 and all $t \in [0,b]$ and $y \in Y$. In fact, suppose that $\bar{x}$ and x are solutions of equation (1.1), then we get

$$d(\bar{x} - x)/dt + A(t,x)(\bar{x} - x) = f(t,\bar{x}) - f(t,x) - [A(t,\bar{x}) - A(t,x)]\bar{x}$$

$0 \leq t \leq b'$, $\bar{x}(0) - x(0) = 0$. Hence, the uniqueness follows if b' is small.

Graff (1979) proved a general uniqueness theorem for the nonlinear Cauchy problem which implies uniqueness of the solution constructed in Theorem 1.1 under more general conditions than the above remark.

2. A SIMPLIFIED APPROACH TO LINEAR EVOLUTION EQUATIONS

The Evolution Operator

Let X, Y, Z be Banach spaces, and denote by $\|\cdot\|_X$, $\|\cdot\|_Y$, $\|\cdot\|_Z$ the norms in X, Y, Z, respectively. We assume that the embedding $Z \subset Y \subset X$ is continuous and Y, Z are dense in X, Y, respectively. For each $t \in [0,b]$ and $y \in Y$, we have $A(t)y \in X$, and the mapping $A(t)$: $Y \to X$ is linear and continuous. Now consider the Cauchy problem for the linear equation

$$Lx(t) \quad \frac{dx}{dt} + A(t)x - f(t) = 0 \qquad 0 \le b \le b \qquad x(0) = y \tag{2.1}$$

Denote by $B(Y,X)$ the space of all bounded linear operators on Y to X with norm denoted by $\|\cdot\|_{Y,X}$. We write $B(X)$ for $B(X,X)$ and $\|\cdot\|_X$ for $\|\cdot\|_{X,X}$.

We denote by $G(X)$ the set of all negative infinitesimal generators of C_0-semigroups $\{U(t)\}$ on X. If $-A \in G(X)$, then $\{U(t)\} = \{e^{-tA}\}$, $0 \le t < \infty$, is the semigroup generated by $-A$.

2.1 DEFINITION (see Kato, 1970) Y ($Y \subset X$) is said to be admissible with respect to A (or A-admissible) if $\{e^{-tA}\}$ leaves Y invariant and forms a semigroup of class C_0 on Y.

We assume, in addition, that for each $t \in [0,b]$ and $z \in Z$, we have $A(t)z \in Y$, and the mapping $A(t)$: $Z \to Y$ is linear and continuous.

2.2 DEFINITION Z ($Z \subset Y \subset X$) is inherently A-admissible if Y is A-admissible and the semigroup of class C_0 on Y which is formed by $\{e^{-tA}\}$ leaves Z invariant and in turn forms a semigroup of class C_0 on Z.

Let $\{A(t)\}$, $0 \le t \le b$, be a family of linear operators from Y to X such that $\{A(t)\} \subset G(X)$.

2.3 DEFINITION (see Kato, 1975) A pair $\{M_X, \beta_X\}$ where $M \ge 1$ and β is a nonnegative function on $[0,b]$ is called an X-stability index for $\{A(t)\}$ if for every finite set of real numbers $a \le t_1 \le \cdots \le t_k \le b$ and arbitrary nonnegative numbers $s_1, \ldots, s_k$, we have

$$\left\| \prod_{j=1}^{k} e^{-s_j A(t_j)} \right\|_X \le M e^{\sum_{j=1}^{k} s_j \beta(t_j)} \tag{2.2}$$

where the product on the left is time ordered; i.e., each term with a given t_j appears on the left of those terms with smaller t_j's. $\{A(t)\}$, $t \in [0,b]$, is called quasistable in X or X-stable if $\beta \in L^1([0,b])$. A similar definition holds for Y-stability; i.e., $\|\cdot\|_{X,X}$ in (1.2) is replaced by $\|\cdot\|_{Y,Y}$ and $\{M_X, \beta_X\}$ is replaced by $\{M_Y, \beta_Y\}$ provided that Y is A(t)-admissible.

Let $A(t) \in G(X)$, $0 \le t \le b$. We make the following assumptions.

(i) $\{A(t)\}$ is quasistable with stability index $\{M_X, \beta_X\}$.

(ii) Y is A(t)-admissible for each $t \in [0,b]$. If $\tilde{A}(t) \in G(Y)$ is the part of A(t) in Y, then $\{\tilde{A}(t)\}$ is quasistable in Y with index $\{M_Y, \beta_Y\}$. Moreover, Z is $\tilde{A}(t)$-admissible (or inherently admissible), and if $\tilde{\tilde{A}}(t) \in G(Z)$ is the part of $\tilde{A}(t)$ in Z, then $\{\tilde{\tilde{A}}(t)\}$ is (inherently) quasistable in Z with index $\{M_Z, \beta_Z\}$.

(iii) $A(t) \in B(Y,X)$ for each $t \in [0,b]$, and the map $t \to A(t)$ is norm-continuous in $B(Y,X)$ · $Z \supset D(\tilde{A}(t))$ for each $t \in [0,b]$, and $\tilde{A}(t) \in B(Z,Y)$. In addition, the map $t \to A(t)$ is norm-continuous in $B(Z,Y)$.

Thus Z is inherently A(t)-admissible.

Following Kato (1970, 1973) and using his notation, we construct the sequence $\{U_n(t,s)\}$ of approximate evolution operators, defined on the triangle Δ: $0 \le s \le t \le b$ with values in X.

We assume that β_X, β_Y, and β_Z are nonnegative and Lebesgue integrable, and then choose partitions $\{I_{nk}\}$ of $I = [0,b]$ and numbers $t_{nk} \in I_{nk}$ in such a way that $|I_{nk}| \to 0$, as $n \to \infty$, and that the corresponding Riemann step functions $\beta_{X,n} = \{I_{nk}, \beta_X(t_{nk})\}$, $\beta_{Y,n} = \{I_{nk}, \beta_Y(t_{nk})\}$, and $\beta_{Z,n} = \{I_{nk}, \beta_Z(t_{nk})\}$ converge to β_X, β_Y, and β_Z, respectively, in L^1-norm as well as pointwise a.e. With this choice of $\{I_{nk}\}$ and $t_{nk} \in I_{nk}$ we construct the step function $A_n = \{I_{nk}, A(t_{nk})\}$

and the associated evolution operator U_n as in Kato (1970, 1973), where $U_n(t,s) = \exp[-(t - s)A(t_{nk})]$ if $t,s \in I_{nk}$ and $s \le t$, and put $U_n(t,r) = U_n(t,s)U_n(s,r)$ if $r \le s \le t$. Then we obtain

$$\|U_n(t,s)\|_X \le M_X \exp\|\beta_X\|_1 \tag{2.3}$$

$$\|U_n(t,s)\|_Y \le M_Y \exp\|\beta_Y\|_1 \tag{2.4}$$

$$\|U_n(t,s)\|_Z \le M_Z \exp\|\beta_Z\|_1 \tag{2.5}$$

where $\|\cdot\|_1$ denotes the L^1-norm and we note that $U_n(t,s)Y \subset Y$ by (ii). In order to show that $\lim U_n(t,s)y$ exists in X for $y \in Y$, uniformly in t,s, we use the following identity:

$$U_n(t,r)y - U_m(t,r)y = -\int_r^t U_n(t,s)(A_n(s) - A_m(s))U_m(s,r)y\, ds$$

which is obtained by differentiating $U_n(t,s)U_m(s,r)y$ in s and integrating. We note that U_n satisfies

$$(d/ds)U_n(t,s)y = U_n(t,s)A_n(s)y \qquad y \in Y$$

$$(d/dt)U_n(t,s)y = -A_n(t)U_n(t,s)y \qquad y \in Y \tag{2.6}$$

Since Y is dense in X, it follows from (2.3) that $\lim U_n(s,t)x = U(t,s)x$ exists for each $x \in X$, uniformly in t,s. Kato's (1970) argument can be continued to show that $U = \lim U_n$ satisfies his Theorem 4.1, and in particular,

$$(d/ds)U(t,s)y = U(t,s)A(s)y \qquad y \in Y \qquad 0 \le s \le t \le b$$

It is easily seen that the same argument can be applied with X,Y replaced by Y,Z, respectively. It follows then that the evolution operator U restricted to Y yields a mapping U: $\Delta \to B(Y)$ which also satisfies Theorem 4.1 (Kato, 1970) and in particular, $U(\cdot,\cdot)y$ is jointly continuous from Δ into Y for $y \in Y$. In fact, we note that $\lim U_n(t,s)z = U(t,s)z$ exists in Y for each $z \in Z$, uniformly in t,s, and that Z is dense in Y by assumption, and use relations (2.4) and (2.5). Then the same argument can be repeated in this case.

2.1 LEMMA Suppose that conditions (i) to (iii) are satisfied, and let $y \in Y$. Then $v(t) = U(t,r)y$ satisfies the equation

$$\frac{dv}{dt} + A(t)v(t) = 0 \qquad 0 \le t \le b \qquad v(r) = y \tag{2.7}$$

Proof: Let $z_n(t) = U_n(t,r)y$. Then we have

$$\frac{dz_n}{dt} = -A_n(t)U_n(t,r)y$$

and $\lim z_n(t) = U(t,r)y$ in Y. Hence,

$$\begin{aligned} Lz_n(t) &\equiv \frac{dz_n}{dt} + A(t)z_n(t) \\ &= [A(t) - A_n(t)]z_n(t) \end{aligned}$$

and consequently,

$$\lim Lz_n(t) = 0 \qquad \text{in } X$$

Since

$$\lim A(t)z_n(t) = A(t)v(t) \qquad \text{in } X$$

the conclusion follows easily. ∎

It follows from Lemma 2.1 that

$$(d/dt)U(t,s)y = -A(t)U(t,s)y$$

where $y \in Y$ is arbitrary.

We are now in a position to prove the following.

2.1 THEOREM Let conditions (i), (ii), and (iii) be satisfied. Then there exists a unique evolution operator $U = \{U(t,s)\}$ such that $U(t,s) \in B(X)$, where $U(t,s)$ is defined on the triangle Δ: $b \ge t \ge s \ge 0$ and has the following properties.

(a) For each $x \in X$, $U(\cdot,\cdot)x$ is jointly continuous from Δ into X, with $U(s,s) = 1$, and $\sup\|U(t,s)\|_X < \infty$.

(b) $U(t,s)U(s,r) = U(t,s)$.

(c) $U(t,s)Y \subset Y$ and $\sup\|U(t,s)\|_Y < \infty$. $U(\cdot,\cdot)y$ is jointly continuous from Δ into Y, for each $y \in Y$.

(d_1) $(d/ds)U(t,s)y = U(t,s)A(s)y \qquad y \in Y \qquad 0 \le s \le t \le b$

(d_2) $(d/dt)U(t,s)y = -A(t)U(t,s)y \qquad y \in Y \qquad 0 \le s \le t \le b$

Proof: It was shown above that $U = \lim U_n$ is an evolution operator with the above properties, and it follows from (2.3) that $\|U(t,s)\|_X \le M_X\|\beta_X\|_1$. Moreover, $U(t,s)Y \subset Y$ and $\|U(t,s)\|_Y \le M_Y\|\beta_Y\|_1$ by (1.4). The uniqueness of U can be proved in the same way as in Kato (1970). In fact, suppose that V(t,s) satisfies (a) and (d_1). Then differentiating $V(t,s)U_n(t,s)y$ in s, using (d_1) and (2.6) and integrating, we obtain for $y \in Y$,

$$V(t,r)y - U_n(t,r)y = \int_r^t V(t,s)(A(s) - A_n(s))U_n(s,r)y\ ds$$

As a consequence of this identity we obtain by virtue of (a) and (2.4) that

$$v(t,y)y = \lim U_n(t,s)y = U(t,r)y$$

Hence, $V(t,r) = U(t,r)$. ∎

2.1 REMARK Let us notice that the conclusion of Kato's (1973) Theorem 1 is stronger than that of Theorem 1.1 in the sense that the continuity in (a) and (c) is strong from Δ into B(X) and B(Y), respectively, and the derivatives in (d_1) and (d_2) exist in the strong sense in B(Y,X) and are strongly continuous from Δ into B(Y,X). However, Kato's assumptions require the existence of a family $S = \{S(t)\}$, $0 \le t \le b$, of isomorphisms of Y onto X, with some properties.

It follows from Kato (1970, 1973) that condition (ii) is implied by the existence of families $\{S_Y(t)\}$, $\{S_Z(t)\}$ of isomorphisms of Y, Z onto X, respectively, such that

$$S_Y(t)A(t)S_Y(t)^{-1} = A(t) + B(t) \qquad B(t) \in B(X)$$

$$S_Z(t)A(t)S_Z(t)^{-1} = A(t) + C(t) \qquad C(t) \in B(X)$$

where $\{S_Y(t)\}$ and $\{S_Z(t)\}$, $\{B(t)\}$, and $\{C(t)\}$ satisfy condition (B) of Kato (1973).

The Inhomogeneous Equation

2.2 THEOREM Suppose that the assumptions of Theorem 2.1 are satisfied. Let $y \in Y$ and let $f(\cdot)$ be continuous from $I = [0,b]$ into Y. Then

$$u(t) = U(t,0)y + \int_0^t U(t,s)f(s)\, ds \tag{2.7}$$

is the solution of (2.1), and

$$\|u\|_{\infty,Y} \le \|U\|_{\infty,Y}(\|y\|_Y + \|f\|_{1,Y})$$
$$\|du/dt\|_{\infty,X} \le \|f\|_{\infty,X} + \|A\|_{\infty,Y,X}(\|y\|_Y + \|f\|_{1,Y})$$

where

$$\|f\|_{\infty,X} = \sup_{0\le t\le b} \|f(t)\|_X$$
$$\|f\|_{1,Y} = \int_0^b \|f(t)\|_Y\, dt \qquad \|U\|_{\infty,Y} = \sup\|U(t,s)\|_Y$$
$$\|A\|_{\infty,YX} = \sup_{0\le t\le b} \|A(t)\|_{Y,X}$$

Proof: It follows from Section 1 that the first term on the right of (2.7) satisfies the homogeneous equation (2.1) with $u(0) = y$. In addition, we have

$$\frac{du}{dt} = -A(t)U(t,0)y + f(t) - \int_0^t A(t)U(t,s)f(s)\, ds \qquad \blacksquare$$

An application to quasilinear evolution equations will be given separately.

3. A GENERAL EXISTENCE THEOREM FOR QUASILINEAR EVOLUTION EQUATIONS

The approach to quasilinear evolution equations (see Section 1) is based on the method of contractor directions. A further refinement of that method is presented here which makes it possible to utilize the simplified approach to linear evolution equations.

It turns out that in order to construct the appropriate contractor directions as required by the proposed method it suffices to solve the corresponding linear equations for the operators generating the approximate evolution operators of Kato (1970; 1973). This is possible because of the flexibility of the method of contractor directions.

Let X, Y, Z be Banach spaces and denote by $\|\cdot\|_X$, $\|\cdot\|_Y$, $\|\cdot\|_Z$ the norms in X, Y, Z, respectively. We assume that the embedding $Z \subset Y \subset X$ is continuous and Z is dense in X. Let W_0 be an open ball in Y with center $x_0 \in Z$ and radius $r > 0$, denote by W its closure in Y, and let Z_0 be an open ball in Z with the same center x_0 and radius $R > 0$. Given $0 < b$, denote by C(0,b;X), C(0,b;Y), C(0,b;Z) the spaces of all continuous functions $x = x(t)$, $y = y(t)$, $z = z(t)$ defined on the interval [0,b] with values in X, Y, Z and norms

$$\|x\|_{\infty,X} = \sup_t [\|x(t)\| : \quad 0 \le t \le b]$$

$$\|y\|_{\infty,Y} = \|y\|_C = \sup_t [\|y(t)\|_Y : \quad 0 \le t \le b]$$

$$\|z\|_{\infty,Z} = \|z\|_D = \sup_t [\|z(t)\|_Z : \quad 0 \le t \le b]$$

respectively. Denote by $C^1(0,b;X)$ the vector space of all continuously differentiable functions from [0,b] into X. In what follows Z may be incomplete.

We assume that there exists a vector space $V \subset Z$ which is dense in $Z(\| \ \|_Z)$ and has the following property. Corresponding to each $t \in [0,b]$ and x contained in the norm $\|\cdot\|_Y$-closure of $V_0 = V \cap Z_0 \cap W_0$, the mapping $A(t,x)$: $Y \to X$ is linear and continuous. We may assume that $\|\cdot\|_Z \ge \|\cdot\|_Y$ and $\|x_0\|_Z < R$. We also assume $x_0 = 0$.

Consider the Cauchy problem for the quasilinear equation

$$Px(t) \equiv \frac{dx}{dt} + A(t,x)x - f(t,x) = 0 \qquad 0 \le t \le b \qquad x(0) = x_0 \tag{3.1}$$

We make the following assumptions.

(A_1) $f(t,v) \in Z$ for all $(t,v) \in [0,b] \times V_0$, and $f(t,x) \in Y$ for all $t \in [0,b]$ and x contained in the $\|\cdot\|_Y$-closure of V_0. If

$(t,x) \in [0,b] \times V_0$, then $A(t,x)v \in z$ for all $v \in V$. We assume that $x_0 \in V_0$ and p_0 is such that

$$\|A(t,x_0)x_0 - f(t,x_0)\|_Y \le p_0 \qquad \text{for all } 0 \le t \le b \tag{3.2}$$

(A_2) Let $x \in C(0,b;V_0(\|\cdot\|_Z)) \cap C^1(0,b;X)$ with $x(0) = x_0$.

There exists a sequence $\{A_m(t,x)\}$ of linear continuous mappings which has the following properties. For each $m = 1, 2, \ldots,$ the equation

$$\frac{dz}{dt} + A_m(t,x)z - y(t) = 0 \qquad 0 \le t \le b \qquad z(0) = x_0 \tag{3.3}$$

where $y \in L^1(0,b;V(\|\cdot\|_Z)) \cap C(0,b;X)$, has a solution $z \in C(0,b;V(\|\cdot\|_Z)) \cap C^1(0,b;X)$ such that

$$\|z\|_D \le bC\|y\|_D \tag{3.4}$$

where C is a global constant which is independent of both n and x. Put $z = x + h$. Then, obviously, h is a solution of the equation

$$\frac{dh}{dt} + A_m(t,x)h + \left[\frac{dx}{dt} + A_n(t,x)x - y\right] = 0 \tag{3.5}$$

$0 \le t \le b$, $h(0) = 0$, and we assume that

$$\|h\|_C \le bC\|\frac{dx}{dt} + A_m(\cdot,x)x - y\|_C \tag{3.6}$$

whenever $\|dx/dt + A(\cdot,x)x - y\|_C < 2p_0$.

Moreover, we assume that

$$\lim\|A_n(\cdot,x) - A(\cdot,x)\|_{\infty,Z,Y} = 0 \tag{3.7}$$

(A_3) There exists a constant K_2 such that

$$\|[A(t,u) - A(t,v)]y\|_X \le K_2\|u - v\|_Y\|y\|_Y \tag{3.8}$$

for all $t \in [0,b]$, and u,v contained in the $\|\cdot\|_Y$-closure of V_0, and $y \in Y$.

(A_4) There exists a constant K_3 such that

$$\|f(t,u) - f(t,v)\|_Y \le K_3\|u - v\|_Y \tag{3.9}$$

for all $t \in [0,b]$ and u,v contained in the $\|\cdot\|_Y$-closure of V_0. We also assume that $f(\cdot,x)$ can be approximated as follows. Let

$$x \in C(0,b;V_0(\|\cdot\|_Z)) \cap C^1(0,b;X)$$

Then for arbitrary $\omega > 0$, there exists a function $y_\omega \in L^1(0,b;V(\|\cdot\|_Z)) \cap C(0,b;X)$ such that

$$\|f(\cdot,x) - y_\omega\|_D < \omega \tag{3.10}$$

and we also assume that d_0 is such that

$$\|f(\cdot,v)\|_D \le d_0 \tag{3.11}$$

for all $v \in V_0$.

Finally, we assume the following extra Lipschitz condition.

(A_5) There exists a constant K_1 such that

$$\|[A(t, v + \varepsilon h) - A(t,v)]u\|_Y \le K_1\varepsilon\|h\|_Y\|u\|_Z \tag{3.12}$$

for all $t \in [0,b]$, $v \in V_0$, $h,u \in V$, and $0 < \varepsilon \le 1$ such that $v + \varepsilon h \in V_0$.

One can weaken this hypothesis by replacing in (3.12) v by $x \in C(0,b;V_0(\|\cdot\|_Z)) \cap C^1(0,b;X)$, $x(0) = x_0$, where $z = x + h$ is a solution of equation (3.3).

Let us mention that if (3.10) can be satisfied with $y_\omega \in C(0,b;V(\|\cdot\|_Z))$, then also in ($A_2$) we assume that $y \in C(0,b;V(\|\cdot\|_Z))$.

3.1 LEMMA If b is sufficiently small, then $\|z - x_0\|_D < R$ whenever z is a solution of equation (3.3), where $x(0) = x_0$ and $x \in C(0,b;V_0(\|\cdot\|_Z)) \cap C^1(0,b;X)$ and y satisfies relation (3.10).

Proof: The proof is obvious, by virtue of (3.4), (3.1), and (3.10) with $\omega < d_0$. ■

3.2 LEMMA There exists a constant K_4 such that

$$\|[A(\cdot, x + \varepsilon h) - A(\cdot,x)](x + \varepsilon h)\|_C \le \varepsilon K_4\|h\|_C \tag{3.13}$$

where $x \in C(0,b;V_0(\|\cdot\|_Z)) \cap C^1(0,b;X)$, $x(0) = x_0$, and $z = x + h$ is a solution of equation (3.3) and $0 < \varepsilon \leq 1$ is such that $x + \varepsilon h$ is contained in W_0.

Proof: The proof is exactly the same as in Lemma 1.2. ∎

Now let $x \in C(0,b;V_0(\|\cdot\|_Z)) \cap C^1(0,b;X)$, $x(0) = x_0$, with Px contained in Y and $\|Px\|_C < \infty$. Let $y \in L^1(0,b;V(\|\cdot\|_Z)) \cap C(0,b;X)$ satisfy (3.10) with small ω to be determined. Let $z = x + h$ be a solution of equation (3.3) with n sufficiently large to be determined. Then we obtain by virtue of (3.6),

$$\begin{aligned}\|h\|_C &\leq bC\|dx/dt + A_n(\cdot,x)x - y\|_C \\ &\leq bC(\|Px\|_C + \|A_n(\cdot,x) - A(\cdot,x)\|_{\infty,Z,Y}\|x\|_Z) \\ &\quad + bC\|y - f(\cdot,x)\|_C \\ &\leq bC_1\|Px\|_C\end{aligned}$$

if $C < C_1$ and ω is sufficiently small provided that n is sufficiently large, by virtue of (3.7). Hence, we get

$$\|h\|_C \leq bC_1\|Px\|_C \tag{3.14}$$

Now put $\bar{x} = x + \varepsilon h$, where $0 < \varepsilon \leq 1$ is so small that $\bar{x}$ is contained in W_0. Thus by virtue of (3.14), it suffices to choose ε so as to satisfy

$$\varepsilon bC_1\|Px\|_C < r - \|x - x_0\|_C \tag{3.15}$$

provided $\|x - x_0\|_C < r$. Furthermore, let $0 < b' \leq b$ and q be such that

$$\bar{q} = b'C_1(K_4 + K_3) < q < 1 \quad \text{and} \quad C < C_1 \tag{3.16}$$

Now we can prove the following.

3.3 LEMMA The following relations hold:

$$\|P(x + \varepsilon h) - (1 - \varepsilon)Px\|_C \leq q\varepsilon\|Px\|_C \tag{3.17}$$

$$\|h\|_C \leq b'C_1\|Px\|_C \tag{3.18}$$

and $\bar{x} = x + \varepsilon h$ satisfies the following relation

$$\bar{x} \in C(0,b';V_0(\|\cdot\|_Z)) \cap C^1(0,b;X) \qquad \bar{x}(0) = x_0 \tag{3.19}$$

and $\|\bar{x} - x_0\|_C < r$ provided that $\omega \le (1/2)(\bar{q} - q)\|Px\|_C$ and n is sufficiently large.

Proof: We have the following identity.

$$\begin{aligned}
&P(x + \varepsilon h) - Px + \varepsilon Px \\
&= [A(\cdot, x + \varepsilon h) - A(\cdot,x)](x + \varepsilon h) \\
&\quad - [f(\cdot, x + \varepsilon h) - f(\cdot,x)] \\
&\quad + [dh/dt + A_n(t,x)h + dx/dt + A_n(\cdot,x)x - y_\omega] \\
&\quad + \varepsilon[A(t,x) - A_n(t,x)]z \\
&\quad + \varepsilon[y_\omega - f(\cdot,x)]
\end{aligned}$$

To obtain an estimate for the first term of the above identity we use relation (3.13) of Lemma 3.2 and (3.14) with b replaced by b'. An estimate for the second term follows from relations (3.9) and (3.14) with b replaced by b'. The third term is equal to zero. The fourth term converges to zero in Y, uniformly in t, by virtue of (3.7) and Lemma 3.1. Hence, n in equation (3.3) can be chosen so large that the fourth term will be dominated by $1/2(q - \bar{q})\|Px\|_C$. An estimate for the fifth term follows from (3.10) with $\omega \le 1/2 \times (q - \bar{q})\|Px\|_C$. Thus we obtain the following estimate by combining the first two terms of the above identity and using (3.14) with b replaced by b'.

$$\begin{aligned}
\|P(x + \varepsilon h) - Px + \varepsilon P\|_C &\le \varepsilon b'C_1(K_4 + K_3)\|Px\|_C \\
&\quad + \varepsilon(q - \bar{q})\|Px\|_C \\
&= \varepsilon\bar{q}\|Px\|_C + \varepsilon(q - \bar{q})\|Px\|_C
\end{aligned}$$

Relation (3.18) results from (3.14) with b replaced by b'. Furthermore, we have

$$\bar{x} = x + \varepsilon h = (1 - \varepsilon)x + \varepsilon z$$

Since $\bar{x}$ is contained in W_0 because so is x by assumption, we conclude from the above that $\bar{x}$ is contained in Z_0. It is now clear that relation (3.19) holds. ■

3.4 LEMMA Let the sequence

$$\{x_n\} \subset C(0,b;V_0(\|\cdot\|_Z)) \cap C^1(0,b;X) \tag{3.20}$$

be such that

$$\|x_n - x\|_C \to 0 \qquad \|Px_n - Px\| \to 0 \qquad \|x - x_0\|_C < r \tag{3.21}$$

Let q', C_1 be such that

$$\bar{q} < q' < q < 1 \qquad \text{and} \qquad b'C_1 < \bar{C}_1 < \bar{C} \tag{3.22}$$

[see (3.16)]. Then there exist $0 < \varepsilon \leq 1$ and

$$\bar{x} \in C(0,b;V_0(\|\cdot\|_Z)) \cap C^1(0,b';X) \tag{3.23}$$

such that

$$\|P\bar{x} - (1 - \varepsilon)Px\|_C \leq \varepsilon q'\|Px\|_C \tag{3.24}$$

$$\|\bar{x} - x\|_C \leq \varepsilon\bar{C}_1\|Px\|_C \tag{3.25}$$

Proof: Choose $0 < \varepsilon \leq 1$ such that

$$\varepsilon b'C_1\|Px\|_C < r - \|x - x_0\|_C \tag{3.26}$$

and n so large that

$$\|x_n - x\|_C < r - \|x - x_0\|_C - \varepsilon b'\|Px\|_C \tag{3.27}$$

Having chosen $0 < \varepsilon \leq 1$, we then move with x_n, Px_n closer to x, Px, respectively, so as to satisfy the following relations:

$$\|Px_n - Px\|_C \leq \varepsilon(q' - \bar{q})\|Px\|_C/3 \tag{3.28}$$

where $\bar{q}$ is given by (3.16), and

$$\|x_n - x\| \leq \varepsilon(\bar{C}_1 - b'C_1)\|Px\|_C \tag{3.29}$$

Let $y_n \in L^1(0,b;V(\|\cdot\|_Z)) \cap C(0,b';X)$ be such that

$$\|y_n - f(\cdot,x_n)\|_D \leq \omega \leq \varepsilon(q' - \bar{q})\|Px\|_C/3 \tag{3.30}$$

Now let $z_n = x_n + h_n$ be a solution of the equation

$$\frac{dz_n}{dt} + A_m(t,x_n)z_n - y_n = 0 \tag{3.31}$$

$0 \le t \le b' \le b$, $z_n(0) = x_0$, with m to be determined. Then we get

$$\frac{dh_n}{dt} + A_m(t,x_n)h_n + \frac{dx_n}{dt} + A_m(\cdot,x_n)x_n - y_n = 0 \tag{3.32}$$

$h_n(0) = 0$, and by virtue of (3.6), we obtain

$$\begin{aligned}\|h_n\|_C &\le b'C\|dx_n/dt + A_m(\cdot,x_n)x_n - y_n\|_C \\ &\le b'C(\|Px_n\|_C + \|y_n - f(\cdot,x_n)\|_C) \\ &\quad + b'C\|[A_m(\cdot,x_n) - A(\cdot,x_n)]x_n\|_C \\ &\le b'C(\|Px\|_C + \|Px_n - Px\|_C + \|y_n - f(\cdot,x_n)\|_C) \\ &\quad + b'C\|[A_m(\cdot,x_n) - A(\cdot,x_n)]x_n\|_C \\ &\le b'C_1\|Px\|_C\end{aligned}$$

Hence, we obtain for $C_1 > C$,

$$\|h_n\|_C \le b'C_1\|Px\|_C \tag{3.33}$$

if m,n are sufficiently large with n satisfying relations (3.27) to (3.30).

In order to prove relation (3.24), the following identity is used.

$$\begin{aligned}P\bar{x} - (1 - \varepsilon)Px &= P(x_n + \varepsilon h_n) - (1 - \varepsilon)Px_n + (1 - \varepsilon)[Px_n - Px] \\ &= [A(\cdot, x_n + \varepsilon h_n) - A(\cdot,x_n)](x_n + \varepsilon h_n) \\ &\quad - [f(\cdot, x_n + \varepsilon h_n) - f(\cdot,x_n)] \\ &\quad + \varepsilon[dh_n/dt + A_m(t,x_n)h_n + dx_n/dt \\ &\quad + A_m(\cdot,x_n)x_n - y_n] \\ &\quad + \varepsilon[A_m(\cdot,x_n) - A(\cdot,x_n)]z_n \\ &\quad + \varepsilon[y_n - f(\cdot,x_n)] + (1 - \varepsilon)[Px_n - Px]\end{aligned}$$

To obtain an estimate for the first term of the above identity we use Lemma 3.2 with $x = x_n$ and $h = h_n$. An estimate for the second term follows from (3.9) and (3.33). The third term is equal to zero. The fourth term (with $m \to \infty$) converges to zero in Y, uniformly in t, by virtue of (3.7) and Lemma 3.1. Hence, after fixing $n = \bar{n}$, m can be chosen so large that the fourth term will be dominated by $(q' - q)\|Px\|_C/3$. Estimates for the last two terms follow from (3.30) and (3.28), respectively. Hence, we obtain the following estimate by combining the first two terms of the above identity.

$$\begin{aligned}\|P(x_n + \varepsilon h_n) - (1 - \varepsilon)Px\|_C &\le \varepsilon b'C_1(K_4 + K_3)\|Px\|_C \\ &\quad + 3\varepsilon(q' - \bar{q})\|Px\|_C/3 \\ &= \varepsilon\bar{q}\|Px\|_C + \varepsilon(q' - \bar{q})\|Px\|_C \\ &= \varepsilon q'\|Px\|_C\end{aligned}$$

Hence, we conclude that if $n = \bar{n}$ is so large that relations (3.27) to (3.30) are satisfied, then relation (3.24) holds. Relation (3.25) results from the following inequality

$$\begin{aligned}\|\bar{x} - x\|_C &\le \|x_n - x\|_C + \varepsilon\|h_n\|_C \\ &\le \varepsilon(\bar{C}_1 - b'C_1)\|Px\|_C + \varepsilon b'C_1\|Px\|_C \\ &\quad \varepsilon\bar{C}_1\|Px\|_C\end{aligned}$$

by virtue of (3.29) and (3.33). Finally, since $\bar{x} = (1 - \varepsilon)x_n + \varepsilon z_n$ with $n = \bar{n}$, relation (3.23) holds, by virtue of (3.20) and Lemma 3.3 with $x = x_n$ and $h = h_n$. ■

3.5 LEMMA Let the sequence

$$\{x_n\} \subset C(0,b';V_0(\|\cdot\|_Z)) \cap C^1(0,b';X)$$

be such that

$$\|x_n - x\|_C \to 0 \qquad \text{and} \qquad \|Px_n - z\|_C \to 0 \qquad \text{as } n \to \infty$$

Then $z = Px$.

Proof: The proof is exactly the same as in Lemma 1.5. ■

The following theorem is a generalization of Theorem 1.1.

3.1 THEOREM In addition to the hypotheses (A_1) to (A_5), suppose that the radius r of the open ball $W_0 \subset Y$ with center $x_0 \in V_0$ satisfies the following relation:

$$r > \bar{r} = (1 - q)^{-1}\bar{C}p_0 \exp(1 - q) \qquad bC < \bar{C}$$

where p_0, C, $0 < q < 1$ are determined by (3.2), (3.6), and (3.16), respectively. Then there exist $0 < b' \leq b$ and a function $x \in C(0,b';W(\|\cdot\|_Y)) \cap C^1(0,b';X)$ satisfying equation (3.1) for all $t \in [0,b']$.

Proof: The proof is exactly the same as that of Theorem 1.1. ■

3.1 REMARK We can make the following extra assumption in (A_2) in order to guarantee that $dx/dt + A_m(\cdot,x)x - y \in L^1(0,b;V(\|\cdot\|_Y)) \cap C(0,b;X)$.

Assume that $x,z \in C(0,b;V_0(\|\cdot\|_Z)) \cap C^1(0,b;X)$ implies $A_m(\cdot,x)z \in L^1(0,b;V(\|\cdot\|_Y)) \cap C(0,b;X)$, for almost all m (see Remark 1.1).

4. AN APPLICATION OF THE SIMPLIFIED APPROACH TO LINEAR EVOLUTION EQUATIONS TO SOLVING QUASILINEAR EVOLUTION EQUATIONS

We are now in a position to apply the simplified approach to linear evolution equations (see Section 2) in order to solve quasilinear evolution equations.

To this end we use Theorem 2.1 and utilize only the results concerning the approximate evolution operators discussed in Section 2.

Using the same notation as in the preceding section, we assume that the Banach spaces Y, Z are dense in X, Y, respectively, and that $V = Z$. Thus $V_0 = Z_0 \cap W_0$.

We make the following assumptions.

(A_0) Corresponding to each $(t,v) \in [0,b] \times V_0$, $A(t,v)Z \subset Y$ holds, and the mappings $A(t,v)\colon\ Y \to X$ and $A(t,v)\colon\ Z \to Y$ are linear and continuous. In addition, for each $v \in V_0$, the function $t \to A(t,v)$ is continuous in the $B(Y,X)$ norm as well as in the $B(Z,Y)$ norm.

(A_2') For each $x \in C(0,b;V_0(\|\cdot\|_Z)) \cap C^1(0,b;X)$, the family $\{A(t,x(t))\}$ satisfies conditions (i) and (ii) of Section 2, i.e., Y is admissible and Z is inherently admissible. Moreover, the family $\{A(t,x(t))\}$ is quasistable in X and inherently quasistable in Y and Z with stability indices $\{M_X,\beta_X\}$, $\{M_Y,\beta_Y\}$, and $\{M_Z,\beta_Z\}$ independent of the function x.

We are now in a position to prove the following.

4.1 THEOREM Let condition (A_0) be satisfied. Then Theorem 1.1 holds with (A_2) replaced by (A_2').

Proof: It suffices to show that condition (A_2) of Theorem 1.1 is satisfied. Let $x \in C(0,b;V_0(\|\cdot\|_Z)) \cap C^1(0,b;X)$ with $x(0) = x_0$. As in Section 2, we construct Kato's step functions

$$A_m = \{I_{mk}, A(t_{mk}, x(t_{mk}))\}$$

for the family $A(t,x)$ as well as the approximate evolution operators $U_m(t,s;x)$. Then the solution of equation (3.3) is given by the formula $z = \int_0^t U_m(t,s;x)y(s)\,ds$. Similarly, the solution of equation (3.5) is given by the formula

$$h = \int_0^t U_m(t,s;x)\,[dx/dt + A(s,x) - y(s)]\,ds$$

Hence, the estimates (3.4) and (3.6) follow provided that $dx/dt \in L^1(0,b;V(\|\cdot\|_Y))$, then also $d\bar{x}/dt \in L^1(0,b;V(\|\cdot\|_Y))$, where $\bar{x} = x + \varepsilon h = (1 - \varepsilon)x + \varepsilon z$ and $z = x + h$ is a solution of equation (3.3). Thus the restriction on dx/dt will be satisfied inductively whenever it appears in the proofs of Theorem 1.1 and the corresponding Lemmas 3.2 to 3.4. Relation (iii), Section 2, holds for $A(t,x(t))$ by virtue of (A_0), (3.8), and (3.12). ∎

4.1 REMARK Theorems 3.1 and 4.1 yield only existence of a solution to the quasilinear evolution equation. However, it is easily seen that a uniqueness theorem can be obtained if we assume the following stronger Lipschitz conditions.

$$\|[A(t,u) - A(t,v)]y\|_X \leq K_1\|u - v\|_X\|y\|_Y$$

and

$$\|f(t,u) - f(t,v)\|_X \leq K_2\|u - v\|_X$$

for some $K_1, K_2 > 0$ and all u,v in the $\|\cdot\|_Y$-closure of V_0 and all $t \in [0,b]$ and $y \in Y$.

Graff (1979) proved a general uniqueness theorem for the nonlinear Cauchy problem, which implies uniqueness of the solutions constructed in Theorems 3.1 and 4.1 under more general conditions than the above remark (see Section 1).

5. AN EXISTENCE THEOREM USING THE EXACT EVOLUTION OPERATOR

Kato's (1975) general existence theorem for quasilinear evolution equations in reflexive Banach spaces is local in time, and so are the theorems obtained in Sections 1 and 3 for nonreflexive Banach spaces. However, it turns out that using the method developed there as well as the simplified approach to linear evolution equations, one can prove an existence theorem for quasilinear evolution equations with simpler time estimate. It was also shown in Section 3 that it suffices to utilize the approximate evolution operators constructed in Section 2 in order to solve the auxiliary linear equation for each iteration step. This is also the case here and we have to take advantage of the corresponding evolution operator while solving the appropriate linear equation at each iteration step.

Let X, Y, Z be Banach spaces and denote by $\|\cdot\|_X$, $\|\cdot\|_Y$, $\|\cdot\|_Z$ the norms in X, Y, Z, respectively. We assume that the embedding $Z \subset Y \subset X$ is continuous and Y,Z are dense in X,Y, respectively. Let W_0 be an open ball in Y with center $x_0 = 0 \in Y$ and radius $r > 0$, and denote by W its closure in Y, and let Z_0 be an open ball in Z with the same center $x_0 = 0$ and radius $R > 0$. Given $0 < b$, denote by $C(0,b;X)$, $C(0,b;Y)$, $C(0,b;Z)$ the spaces of all continuous functions $x = x(t)$, $y = y(t)$, $z = z(t)$ defined on the interval $[0,b]$ with values in X, Y, Z, and norms

$$\|x\|_{\infty,X} = \sup_t [\|x(t)\|_X: \quad 0 \le t \le b]$$

$$\|y\|_C = \|y\|_{\infty,Y} = \sup_t [\|y(t)\|_Y: \quad 0 \le t \le b]$$

$$\|z\|_D = \|z\|_{\infty,Z} = \sup_t [\|z(t)\|_Z: \quad 0 \le t \le b]$$

respectively. Denote by $C^1(0,b;X)$ the vector space of all continuously differentiable functions from $[0,b]$ into X, and by V_1 the $\|\cdot\|_Y$-closure of $V_0 = Z_0 \cap W_0$.

Corresponding to each $t \in [0,b]$ and x contained in V_1, the mapping $A(t,x)$: $Y \to X$ is linear and continuous. We may assume that $\|\cdot\|_Z \ge \|\cdot\|_Y$.

Consider the Cauchy problem for the quasilinear equation

$$Px(t) \equiv \frac{dx}{dt} + A(t,x)x - f(t,x) = 0 \qquad (5.1)$$
$$0 \le t \le b \qquad x(0) = x_0$$

We make the following assumptions.

(A_1) $f(t,v) \in Z$ for all $(t,v) \in [0,b] \times V_0$, and $f(t,x) \in Y$ for all $t \in [0,b]$ and x contained in V_1. We assume that $p_0 > 0$, d_0 and K_3 are such that

$$\|f(t,0)\|_Y \le p_0 \qquad \text{for all } 0 \le t \le b \qquad (5.2)$$

$$\|f(\cdot,v)\|_D \le d_0 \qquad (5.3)$$

$$\|f(t,u) - f(t,v)\|_Y \le K_3\|u - v\|_Y \qquad (5.4)$$

for all $t \in [0,b]$ and u,v contained in V_1, and $f(\cdot,0)$ is continuous in Y.

(A_2) Corresponding to each $(t,v) \in [0,b] \times V_0$, $A(t,v)Z \subset Y$, and the linear mapping $A(t,v)$: $Z \to Y$ is continuous. In addition, for each $v \in V_0$, the function $t \to A(t,v)$ is continuous in the $B(Y,X)$

norm as well as in the B(Z,Y) norm, where B(Y,X) and B(Z,Y) are the Banach spaces of all linear continuous operators from Y,Z into X,Y with uniform norms $\|\cdot\|_{Y,X}$, $\|\cdot\|_{Z,Y}$ on B(Y,X), B(Z,Y), respectively.

Denote by G(X) the set of all negative infinitesimal generators of C_0-semigroups U(t) on X. If $-A \in G(X)$, then $U(t) = e^{-tA}$, $0 \le t \le \infty$, is the semigroup generated by $-A$.

(A_3) $A(t,x) \in G(X)$ for each $(t,x) \in [0,b] \times V_0$. For each $x \in C(0,b;V_0(\|\cdot\|_Z)) \cap C^1(0,b;X)$, the family $\{A(t,x(t))\}$ satisfies conditions (i) and (ii) of Y is admissible and Z is (inherently) admissible. Moreover, the family $\{A(t,x(t))\}$ is quasistable in X and inherently quasistable in Y and Z. We also assume that the corresponding evolution operators have the following property. There exists an integrable nonnegative function β defined on [0,b] such that

$$\|U(t,s;x)\|_{Y,Y} \le \beta(s) \qquad 0 \le s \le t \le b \tag{5.5}$$

for all $t \in [0,b]$ and $x \in C(0,b;V_0(\|\cdot\|_Z)) \cap C^1(0,b;X)$.

(A_4) For each $x \in C(0,b;V_0(\|\cdot\|_Z)) \cap C^1(0,b;X)$, the equation

$$\frac{dz}{dt} + A(t,x)z - f(t,x) = 0 \qquad 0 \le t \le b \qquad z(0) = x_0 \tag{5.6}$$

has a solution $z \in C(0,b;Z(\|\cdot\|_Z)) \cap C^1(0,b;X)$ such that

$$\|z - x_0\|_D < R \tag{5.7}$$

Now put $z = x + h$. Then obviously, h is a solution of the equation

$$\frac{dh}{dt} + A(t,x)h + [dx/dt + A(t,x)x - f(t,x)] = 0 \tag{5.8}$$

$0 \le t \le b$, $h(0) = 0$, and we have

$$h(t) = \int_0^t U(t,s;x)[dx/ds + A(s,x)x - f(s,x)]\,ds \tag{5.9}$$

(A_5) There exist constants K_1 and K_2 such that

$$\|[A(t,u) - A(t,v)]y\|_X \le K_2\|u - v\|_Y\|y\|_Y \tag{5.10}$$

for all $t \in [0,b]$ and u,v contained in V_1, and

$$\|[A(t,z) - A(t,x)]z\|_Y \le K_1\|h\|_Y\|z\|_Z \tag{5.11}$$

for all $t \in [0,b]$, $x \in V_0$, and $z \in V_0$, where $h = z - x$.

5.1 LEMMA There exists a constant K_4 such that

$$\|[A(\cdot, x + h) - A(\cdot,x)](x + h)\|_C \le K_4\|h\|_C \tag{5.12}$$

where $x \in C(0,b;V_0(\|\cdot\|_Z)) \cap C^1(0,b;X)$, and $z = x + h$ is a solution of equation (1.6).

Proof: The proof follows immediately with $K_4 = K_1R$ as a consequence of (5.7) and (5.11). ■

Now let $x \in C(0,b;V_0(\|\cdot\|_Z)) \cap C^1(0,b;X)$, $x(0) = x_0$, with Px contained in Y. Let $z = x + h$ be a solution of equation (5.6). The following lemma holds.

5.2 LEMMA Suppose that $z = x + h$ is a solution of equation (5.6). Then the following inequality holds.

$$\|P(x + h)(t)\|_Y \le K \int_0^t \beta(s)\|Px(s)\|_Y\, ds \tag{5.13}$$

where $K = K_4 + K_3$.

Proof: We have the following identity.

$$\begin{aligned} P(x + h)(t) = {} & [A(t,s(t) + h(t)) - A(t,x(t))](x(t) + h(t)) \\ & - [f(t,x(t) + h(t)) - f(t,x(t))] \\ & + [dh/dt + A(t,x)h + dx/dt + A(t,x)x - f(t,x)] \end{aligned}$$

Hence, we obtain by (5.8), (5.12), and (5.4) that

$$\|P(x + h)(t)\|_Y \le (K_4 + K_3)\|h(t)\|_Y$$

and relation (5.13) follows from (5.9) and (5.5). ■

5.3 LEMMA Let the sequence

$$\{x_n\} \subset C(0,b;V_0(\|\cdot\|_Z)) \cap C^1(0,b;X)$$

be such that

$$\|x_n - x\|_C \to 0 \qquad \text{and} \qquad \|Px_n - z\|_C \to 0 \qquad \text{as } n \to \infty$$

Then $z = Px$.

Proof: The proof is exactly the same as in Lemma 1.5. ∎

We are now in a position to prove the following.

5.1 THEOREM In addition to the hypotheses (A_1) to (A_5), suppose that the solution z of equation (5.6) satisfies

$$\|z - x_0\| < r \tag{5.14}$$

Then there exists a function

$$x \in C(0,b;W(\|\cdot\|_Y)) \cap C^1(0,b;X)$$

satisfying equation (5.1) for all $t \in [0,b]$.

Proof: We construct a sequence

$$\{x_n\} \subset C(0,b;V_0(\|\cdot\|_Z)) \cap C^1(0,b;X)$$

as follows. Start with x_0 and suppose that $x_1, \ldots, x_n$ are known. Next we solve the equation

$$\frac{dz}{dt} + A(t,x_n)z - f(t,x_n) = 0 \qquad 0 \le t \le b \qquad z(0) = x_0 \tag{*}$$

and let $z_n = x_n + h_n$ be its solution. Then we obtain from relation (5.13) with $x = x_n$ and $h = h_n$ that for $x_{n+1} = z_n$,

$$\|Px_{n+1}(t)\|_Y \le K \int_0^t \beta(s)\|Px_n(s)\|_Y\, ds$$

Hence, by induction, we get

$$\|Px_{n+1}(t)\|_Y \le \|PO\|_C K^n \int \cdots \int_{0 \le s_1 \le \cdots \le s_n \le t} \beta(s_1)\cdots\beta(s_n)\,ds_1\cdots ds_n$$

$$\le \|PO\|_C K^n \frac{1}{n!} \left[\int_0^t \beta(s)\, ds\right]^n$$

and consequently,

$$\|Px_n\|_C \leq p_0 K^n \frac{1}{n!} [\int_0^b \beta(s)\, ds]^n \tag{5.15}$$

Hence, it follows from (5.9) with $x = x_n$ and $h = h_n$ that

$$\|h_n\|_C \leq p_0 K^n \frac{1}{n!} [\int_0^b \beta(s)\, ds]^{n+1} \tag{5.16}$$

Since $x_n = \Sigma_{i=0}^{n-1} h_i$, relations (5.15) and (5.16) imply that the sequence $\{x_n\}$ is convergent in the $\|\cdot\|_C$ norm to some x and $Px = 0$, by Lemma 5.3. It is clear from (5.14) with $z = x_n$ that $\|x\|_C \leq r$, and the proof is complete. ■

6. AN EXISTENCE THEOREM USING THE APPROXIMATE EVOLUTION OPERATOR

The sequence $\{x_n\}$ constructed in the proof of Theorem 5.1 is obtained by solving equation (*) and using the corresponding evolution operator $U(t,s;x_n)$. We now replace the evolution operator $U(t,s;x_n)$ by its approximation $U_m(t,s,x_n)$ as well as equation (*) by the following:

$$\frac{dz}{dt} + A_m(t,x_n)z - y_n = 0 \qquad 0 \leq t \leq b \qquad z(0) = x_0 \tag{6.1}$$

where Kato's step function $A_m = \{I_{mk}, A(t_{mk})\}$ and the corresponding approximate evolution operator U_m are the same as in Section 2. To do so we make the following changes in the hypotheses of Theorem 1.1.

(A_1') $f(t,v) \in Y$ for all $(t,v) \in [0,b]$ and $v \in V_1$ with $R = r$ from now on.

For arbitrary $x \in C(0,b;V_0(\|\cdot\|_Z)) \cap C^1(0,b;X)$ and $\omega > 0$, there exists a function $y \in C(0,b;Y)$ such that

$$\|f(t,x(t)) - y(t)\|_Z \leq \omega \qquad \text{for all } t \in [0,b] \tag{6.2}$$

The remaining conditions of (A_1) are the same.

(A_2') The conditions are the same as in (A_2).

(A_3') Condition (A_3) holds with (5.5) replaced by the following relations. There exist nonnegative integrable functions $\tilde{\beta}$ and $\tilde{\tilde{\beta}}$ such that

$$\|U_m(t,s,x)\|_{Y,Y} \le \tilde{\beta}(s) \qquad 0 \le s \le t \le b \tag{6.3}$$

and

$$\|U_m(t,s,x)\|_{Z,Z} \le \tilde{\tilde{\beta}}(s) \qquad 0 \le s \le t \le b \tag{6.4}$$

where $\tilde{\beta}$ and $\tilde{\tilde{\beta}}$ are independent of both m and $x \in C(0,b;V_0(\|\cdot\|_Z)) \cap C^1(0,b;X)$.

Then for each $x \in C(0,b;V_0(\|\cdot\|_Z)) \cap C^1(0,b;X)$, and arbitrary m, the equation

$$\frac{dz}{dt} + A_m(t,x)z - y = 0 \qquad 0 \le t \le b \qquad z(0) = x_0 \tag{6.5}$$

has a solution $z \in C(0,b;Z(\|\cdot\|_Z)) \cap C^1(0,b;X)$.

(A_4') We assume that r is such that

$$\|z\|_D \le r \tag{6.6}$$

whenever z is a solution of (6.5).

Now put $z = x + h$. Then obviously, h is the solution of the equation

$$\frac{dh}{dt} + A_m(t,x)h + [dx/dt + A_m(t,x)x - y] \tag{6.7}$$

$0 \le t \le b$, $h(0) = 0$, and we have

$$h(t) = -\int_0^t U_m(t,s;x)\,[dx/ds + A_m(s,x)x - y(s)]\,ds \tag{6.8}$$

(A_5') Conditions are the same as in (A_5).

6.1 LEMMA Given $\varepsilon > 0$, let $y \in C(0,b;Z(\|\cdot\|_Z))$ be such that

$$\|f(\cdot,x) - y\|_D \le \frac{\varepsilon}{2} \tag{6.9}$$

where $x \in C(0,b;V_0(\|\cdot\|_Z)) \cap C^1(0,b;X)$, and let m be so large that

$$\|A(\cdot,x) - A_m(\cdot,x)\|_{\infty,Y,Z} \le \frac{\varepsilon}{2r} \tag{6.10}$$

For $z = x + h$, where z is the solution of equation (6.5), the following relations hold:

$$\|P(x + h)(t)\|_Y \le K \int_0^t \beta(s)\|Px(s)\|_Y\, ds + \varepsilon K \int_0^t \beta(s)\, ds + \varepsilon \tag{6.11}$$

and

$$\|h(t)\|_Y \le \int_0^t \beta(s)\|Px(s)\|_Y\, ds + \varepsilon \int_0^t \beta(s)\, ds$$

Proof: We have the following identity:

$$\begin{aligned} P(x + h) = {} & [A(\cdot, x + h) - A(\cdot,x)](x + h) \\ & - [f(\cdot, x + h) - f(\cdot,x)] \\ & + [dh/dt + A_m(t,x)h + dx/dt + A_m(\cdot,x)x - y] \\ & + [A(t,x) - A_m(t,x)z] \\ & + [y - f(\cdot,x)] \end{aligned}$$

To obtain an estimate for the first term of the above identity we use relations (5.11) and (6.6). An estimate for the second term follows from (5.4). The third term is equal to zero. An estimate for the fourth term results from (6.10). An estimate for the fifth term follows from (6.9). Hence, we obtain by combining the first two terms,

$$\|P(x + h)(t)\|_Y \le K\|h(t)\|_Y + \varepsilon$$

with $K = K_1 r + K_3$. On the other hand, we get from (6.8) and (6.3) with $\tilde{\beta}$ replaced by β that

$$\|h(t)\|_Y \le \int_0^t \beta(s)\|Px(s)\|_Y\, ds + \varepsilon \int_0^t \beta(s)\, ds$$

by virtue of (6.9) and (6.10), and the proof of the lemma is complete. ∎

We are now in a position to prove the following.

6.1 THEOREM Suppose that the hypotheses (A_1') to (A_5') are satisfied. Then there exists a function

$$x \in C(0,b;W(\|\cdot\|_Y)) \cap C^1(0,b;X)$$

satisfying the equation

$$\frac{dx}{dt} + A(t,x)x - f(t,x) = 0 \qquad x(0) = x_0$$

for all $0 \le t \le b$.

Proof: We first choose a nonnegative convergent series

$$\sum_{i=0}^{\infty} \varepsilon_i < \infty \tag{6.13}$$

Next we construct a sequence

$$\{x_n\} \subset C(0,b;V_0(\|\cdot\|_Z)) \cap C^1(0,b;X)$$

as follows. Put $x_0 = 0$, and suppose that $x_1, \ldots, x_n$ are known. Now we solve the equation

$$\frac{dz}{dt} + A_m(t,x_n)z - y_n = 0 \qquad 0 \le t \le b \qquad z(0) = x_0 \tag{6.14}$$

with m to be determined later, and let $z_n = x_n + h_n$ be the solution, and put $x_{n+1} = z_n$. By virtue of Lemma 6.1, we obtain from relations (6.11) and (6.12) with $x = x_n$, $h = h_n$, $\varepsilon = \varepsilon_n$, that

$$\|Px_{n+1}(t)\|_Y \le K \int_0^t \beta(s)\|Px_n(s)\|_Y\, ds + \varepsilon_n K \int_0^t \beta(s)\, ds + \varepsilon_n \tag{6.15}$$

and

$$\|h_n(t)\|_Y \le \int_0^t \beta(s)\|Px_n(s)\|_Y\, ds + \varepsilon_n \int_0^t \beta(s)\, ds \tag{6.16}$$

In order to simplify the computations put $\bar{\varepsilon}_n = (KC + 1)\varepsilon_n$, where $C = \int_0^b \beta(s)\, ds$. Then using the same symbol, ε_n, instead of $\bar{\varepsilon}_n$, we can replace relations (6.15) and (6.16) by the following:

$$\|Px_{n+1}(t)\|_Y \le K \int_0^t \beta(s)\|Px_n(s)\|_Y\, ds + \varepsilon_n \tag{6.17}$$

and

$$\|h_n(t)\|_Y \le \int_0^t \beta(s)\|P_n(s)\|_Y\, ds + \varepsilon_n \tag{6.18}$$

Hence, we obtain from (6.17) by induction,

$$
\begin{aligned}
&\|Px_{n+1}(t)\|_Y \\
&\le K^n \|Px_0\|_C \int_{0\le s_1\le\cdots\le s_n\le t}\!\!\cdots\!\!\int \beta(x_1)\cdots\beta(x_n)\,ds_1\cdots ds_n \\
&+ K^{n-1}\varepsilon_0 \int_{0\le s_1\le\cdots\le s_{n-1}\le t}\!\!\cdots\!\!\int \beta(s_1)\cdots\beta(s_{n-1})\,ds_1\cdots ds_{n-1} \\
&+ K^{n-2}\varepsilon_1 \int_{0\le s_1\le\cdots\le s_{n-2}\le t}\!\!\cdots\!\!\int \beta(s_1)\cdots\beta(s_{n-2})\,ds_1\cdots ds_{n-2} \\
&+ \cdots + K\varepsilon_{n-1} \int_0^t \beta(s_1)\,ds_1 + \varepsilon_{n-1} \qquad (6.19)
\end{aligned}
$$

Hence, we get by (5.2),

$$
\begin{aligned}
\|Px_n\|_C \le p_0(KC)^n \frac{1}{n!} &+ \varepsilon_0(KC)^{n-1}\frac{1}{(n-1)!} \\
&+ \varepsilon_1(KC)^{n-2}\frac{1}{(n-2)!} + \cdots + \varepsilon_{n-2}(KC) + \varepsilon_{n-1} \qquad (6.20)
\end{aligned}
$$

It follows from (6.20) that

$$\sum_{n=0}^{\infty} \|Px_n\|_C \le p_0 \exp(KC) + \exp(KC) \sum_{n=0}^{\infty} \varepsilon_n < \infty$$

Since $x_n = \Sigma_{i=0}^{n-1} h_i$, relations (6.18) and (6.21) imply that the sequence $\{x_n\}$ is convergent in the $\|\cdot\|_C$ norm to some x, and $Px = 0$ by Lemma 5.3. It is clear from (6.6) with $z = x_n$ that $\|x\|_C \le r$, and the proof is complete. ∎

For other results on quasilinear evolution equations in nonreflexive Banach spaces see Kato (1982), Nakata (1983), and references therein.

APPENDIX 1
Iterative Methods of Contractor Directions

1. INTRODUCTION

The general theory of contractor directions presented in Altman (1977) and subsequently developed in a number of investigations (see Altman, 1978a,b; 1979a,b) has proved an extremely useful and powerful tool in the study of nonlinear operator equations in Banach spaces. This theory provides a general method of solving equations under rather weak hypotheses. This fact is due to the structure of the general theory, which has all the advantages of abstract differential calculus, but no differentiation of any kind is involved in the general definition of contractor directions.

The method of contractor directions yields very general local and global existence theorems for nonlinear operator equations in Banach spaces. The theory of contractor directions also provides applications to nonlinear differential and integral equations and optimization theory. An application to the theory of monotone operators is given in Altman (1978b).

The method of contractor directions developed in Altman (1977) requires an elaborate transfinite induction technique. For that reason the method is not constructive, and there was a need for developing a constructive version of the method in order to find solutions of equations in a practically effective way. The first attempt has already been made (see Altman, 1979a). Among other things, the Newton-Kantorovič method with small steps is investigated there, and convergence theorems are proved under less restrictive hypotheses.

It is the purpose of this appendix to present a general theory of iterative methods of contractor directions. One characteristic and natural feature of these methods is the variable steplength. This gives the method a certain flexibility that results in convergence under weaker hypotheses and also in cases where the initial approximate solution is not as good. The theory opens a new direction in the development of iterative methods.

2. THE FUNDAMENTAL LEMMA FOR ITERATIVE METHODS OF CONTRACTOR DIRECTIONS

Let $P: D(P) \subset X \to Y$ be an arbitrary nonlinear operator, and $U_0 = D(P) \cap S$, S being an open ball with center $x_0 \in D(P)$ and radius r. Suppose $\Gamma_x(P)$ has the (B,1)-property with $B \in \mathbb{B}$ (see Section 7), and that $-Px \in \Gamma_x(P)$ for all $x \in U_0$. This means that relations (7.1) and (7.2) are satisfied with $y = -Px$ and $g(x) = 1$ for all $x \in U_0$. The radius r is supposed to satisfy the following relation.

$$r \geq (1 - q)^{-1} \int_0^a s^{-1}B(s)\, ds \quad \text{and} \quad \|Px_0\| \exp(1 - q) \leq a \qquad \text{(c)}$$

We can construct the following algorithm for the method of contractor directions. Given x_0 and $0 < q < 1$, suppose that $x_1, \ldots, x_n$ are already defined. Then we put

$$x_{n+1} = x_n + \varepsilon_n h_n \tag{2.1}$$

where $0 < \varepsilon_n \leq 1$ and h_n are chosen so as to satisfy

$$\|P(x_n + \varepsilon_n h_n) - (1 - \varepsilon_n)Px_n\| \leq q\varepsilon_n\|Px_n\| \tag{2.2}$$

and

$$\|h_n\| \leq B(\|Px_n\|) \tag{2.3}$$

Such a choice is possible on the basis of the assumptions made above. Thus, the sequence $\{x_n\}$ is well defined and we have the following.

2.1 LEMMA The sequence $\{Px_n\}$ is a Cauchy sequence. The sequence $\{x_n\}$ lies in U_0 and converges to an element x^* such that

$$\|x_n - x^*\| \le (1 - q)^{-1} \int_b^{b_n} s^{-1}B(s)\, ds \tag{2.4}$$

for n = 0, 1, ..., where

$$b_n = \|Px_0\| \exp((1 - q)(1 - t_n))$$

$$b = \|Px_0\| \exp((1 - q)(1 - T)) \qquad t_0 = 0$$

$$t_n = \sum_{i=0}^{n-1} \varepsilon_i \qquad \text{and} \qquad T = \sum_{i=0}^{\infty} \varepsilon_i$$

Proof: Consider the sequence $\{a_n\}$ defined as follows.

$$a_0 = \|Px_0\| \qquad a_{n+1} = (1 - (1 - q)\varepsilon_n)a_n \qquad \text{for } n = 1, 2, \ldots$$

By virtue of (2.2), we can prove by induction that

$$\|Px_{n+1}\| \le (1 - (1 - q)\varepsilon_n)\|Px_n\| \le a_{n+1} \qquad \text{for all } n \tag{2.5}$$

In order to prove that $\{Px_n\}$ is a Cauchy sequence, we observe from (2.2) and (2.1) that

$$\|Px_{n+1} - Px_n\| \le (1 + q)\varepsilon_n\|Px_n\| \qquad \text{for all } n \tag{2.6}$$

It results from (2.5) and (2.6) that the series $\Sigma_{n=0}^{\infty} \|Px_{n+1} - Px_n\|$ is dominated by the series $\Sigma_{n=0}^{\infty} \varepsilon_n a_n$, which is convergent by virtue of Remark 7.1 of this appendix, where we can put $B(s) = s$. It follows from the same remark and from relations (2.3) and (2.5) that

$$\sum_{i=n}^{\infty} \varepsilon_i\|h_i\| \le (1 - q)^{-1} \int_b^{b_n} s^{-1}B(s)\, ds \tag{2.7}$$

for n = 0, 1, ..., where b and b_n are the same as in (2.4). The convergence of $\{x_n\}$ and the estimate (2.4) are immediate consequences of relation (2.7). It is easily seen from relations (2.1) and (2.7) with $n = 0$ that the sequence $\{x_n\}$ lies in U_0 if r is subject to condition (c), since $b_0 = \|Px_0\| \exp(1 - q)$, by virtue of (2.4).

3. ITERATIVE METHODS OF CONTRACTOR DIRECTIONS

Let $P: D(P) \subset X \to Y$ be a nonlinear mapping, and put $U_0 = D(P) \cap S$, where S is an open ball with center $x_0 \in D(P)$ and radius r to be defined below.

3.1 DEFINITION A mapping $h: D(P) \to X$ is called a locally uniform strategy for P if

$$\|P(x_n + \varepsilon_n h(x_n)) - (1 - \varepsilon_n)Px_n\|/\varepsilon_n \to 0 \qquad \text{as } \varepsilon_n \to 0 \tag{3.1}$$

whenever $\{x_n\}$ is a Cauchy sequence which lies in U_0 and $\{h(x_n)\}$ is bounded. If, in addition, we have

$$\|h(x)\| \le B(\|Px\|) \tag{3.2}$$

for all $x \in U_0$, where $B \in \mathbb{B}$, then h is called a uniform B-strategy for P.

3.1 LEMMA Suppose that the Fréchet derivative $P'(x)$ exists for all $x \in U = D(P) \cap \bar{S}$, where $\bar{S}$ is the closure of S, and is continuous in U, and has the following property. For arbitrary $x \in U_0$, there exists an element $h(x)$ such that

$$P'(x)h(x) = -Px \tag{3.3}$$

Then the mapping $h = h(x)$ is a locally uniform strategy for P. If, in addition, $h(x)$ satisfies condition (3.2), then the strategy $h = h(x)$ is a locally uniform B-strategy for P.

Proof: Let $\{x_n\}$ be an arbitrary Cauchy sequence in U_0 and let $x^* \in U$ be its limit, and let $P'(x_n)h(x_n) = -Px_n$, for $n = 0, 1, \ldots$. Then we have

$$\|P(x_n + \varepsilon_n h(x_n) - (1 - \varepsilon_n)Px_n\|$$
$$= \|P(x_n + \varepsilon_n h(x_n)) - Px_n - \varepsilon_n P'(x_n)h(x_n)\|$$

$$\leq \varepsilon_n \|h(x_n)\| \int_0^1 \|P'(x_n + \varepsilon_n th(x_n) - P'(x_n)\| \, dt$$

Hence, we obtain

$$\|P(x_n + \varepsilon_n h(x_n)) - (1 - \varepsilon_n)Px_n\|/\varepsilon$$

$$\leq \|h(x_n)\| \int_0^1 \|P'(x_n + \varepsilon_n th(x_n)) - P'(x_n)\| \, dt$$

But

$$\|P'(x_n + \varepsilon_n th(x_n)) - P'(x_n)\|$$

$$\leq \|P'(x_n + \varepsilon_n th(x_n)) - P'(x^*)\| + \|P'(x^*) - P'(x_n)\|$$

and the sequence $\{h(x_n)\}$ is bounded. Moreover, the sequence $\{x_n\}$ converges to x^* and the sequence $\{\varepsilon_n\}$ converges to 0. Hence, it follows that

$$\|P'(x_n + \varepsilon_n th(x_n)) - P'(x_n)\| \to 0 \qquad \text{as } n \to \infty$$

where the convergence is uniform in $0 \leq t \leq 1$. Consequently, we obtain from (3.4) that relation (3.1) is satisfied, and the proof is completed. ∎

The above lemma shows that the existence of a locally uniform strategy requires only continuity of $P'(x)$.

Suppose that P is continuous and has a uniform strategy $h = h(x)$. Then we can define the following algorithm. Given $x_0 \in D(P)$ and arbitrary $0 < \beta < 1$ and $0 < q < 1$, put

$$\Phi(\varepsilon,h,x) = \|P(x + \varepsilon h) - (1 - \varepsilon)Px\|/\varepsilon \qquad 0 < \varepsilon \leq 1$$

If $h_0 = h(x_0)$ and $\Phi(1,h_0,x_0) \leq q\|Px_0\|$, then we put $\varepsilon_0 = 1$ and $x_1 = x_0 + h_0$. If $\Phi(1,h_0,x_0) > q\|Px_0\|$, then there exists an $0 < \varepsilon_0 < 1$ such that

$$\beta q\|Px_0\| \leq \Phi(\varepsilon_0,h_0,x_0) \leq q\|Px_0\|$$

and we put $x_1 = x_0 + \varepsilon_0 h_0$. Assume that $x_0, x_1, \ldots, x_n$ are already defined. In order to define x_{n+1}, we put $\varepsilon_n = 1$ if $\Phi(1,h_n,x_n) > q\|Px_n\|$, and $x_{n+1} = x_n + h_n$, where $h_n = h(x_n)$. If $\Phi(1,h_n,x_n) > q\|Px_n\|$, then there exists a positive $\varepsilon_n < 1$ such that

$$\beta q\|Px_n\| \leq \Phi(\varepsilon_n,h_n,x_n) \leq q\|Px_n\| \tag{3.5}$$

and we put

$$x_{n+1} = x_n + \varepsilon_n h_n \tag{3.6}$$

3.1 THEOREM If P is continuous and has a locally uniform B-strategy in U_0 with

$$r \leq (1 - q)^{-1} \int_0^a s^{-1}B(s)\, ds \quad \text{and} \quad \exp(1 - q)\|Px_0\| \leq a \tag{3.7}$$

then the sequence of iterates x_n lies in U_0 and converges to a solution x^* of equation $Px = 0$. The following error estimate holds.

$$\|x_n - x^*\| \leq (1 - q)^{-1} \int_0^{b_n} s^{-1}B(s)\, ds \tag{3.8}$$

where

$$b_n = \|Px_0\| \exp((1 - q)(1 - t_n))$$
$$b = \|Px_0\| \exp((1 - q)(1 - T))$$

$$t_0 = 0 \qquad t_n = \sum_{i=0}^{n-1} \varepsilon_i \qquad T = \sum_{i=0}^{\infty} \varepsilon_i$$

Proof: Relation (3.5) implies that condition (2.2) is satisfied. It follows from the definition of the B-strategy that also relation (2.3) is fulfilled. Therefore, the hypotheses of Lemma 2.1 are satisfied. Hence, it follows that the sequence $\{x_n\}$ defined by relation (3.6) is a Cauchy sequence and lies in U_0 with r determined by relation (3.7). Since the estimate (3.8) results from (2.4), it remains to show that $Px_n \to 0$ as $n \to 0$. To prove this, we suppose that

(a) The choice $\varepsilon_n = 1$ occurs for an infinite number of n. Then obviously $T = \infty$, and by Remark 7.1, the sequence $\{a_n\}$ is convergent to 0, and so is $\{\|Px_n\|\}$.

If (a) is not the case, then relation (3.5) applies for almost all n. Since the sequence of $\|h_n\| = \|h(x_n)\| \le B(\|Px_n\|)$ is bounded, we have, by virtue of (3.1), that $\Phi(\varepsilon_n, h_n, x_n) \to 0$ if $\varepsilon_n \to 0$ as $n \to \infty$, by virtue of (3.5). On the other hand, if $\{\varepsilon_n\}$ is not convergent to 0, then evidently $T = \infty$. In this case we use the same argument again as in (a) and conclude that $Px_n \to 0$ as $n \to \infty$. This completes the proof. ∎

3.2 THEOREM Suppose that the Fréchet derivative $P'(x)$ is continuous in U with r defined by (3.7), and for every $x \in U_0$, there exists an element $h = h(x)$ satisfying (3.2) and such that $P'(x)h(x) = -Px$. Then the sequence of iterates x_n defined by (3.6) with $h_n = h(x_n)$ lies in U_0 and converges to a solution x^* of equation $Px = 0$. The error estimate (3.8) is also valid.

Proof: The proof follows immediately from Lemma 3.1 and Theorem 3.1. ∎

3.1 REMARK It follows from the error estimate (3.8) that the larger each stepsize ε_n the faster the convergence of the sequence $\{x_n\}$ in both Theorems 3.1 and 3.2. Therefore, the best choice for the stepsize is the largest number $\varepsilon_n \le 1$ which satisfies relation (3.5).

Suppose now that $P'(x)^{-1}$, the inverse of the Fréchet derivative $P'(x)$, exists for all $x \in U_0$ and satisfies the following condition, for some $B \in \mathbb{B}$.

$$\|P'(x)^{-1}Px\| \le B(\|Px\|) \qquad \text{for all } x \in U_0 \tag{3.9}$$

Then we define the following algorithm.

$$x_{n+1} = x_n - \varepsilon_n P'(x_n)^{-1} Px_n \qquad \text{for } n = 0, 1, \ldots \tag{3.10}$$

where $\varepsilon_n = 1$ if $\Phi(1, h_n, x_n) \le 1$ or satisfies relation (3.5) otherwise. Then the following theorem holds for the damped Newton-Kantorovič method (3.10).

3.3 THEOREM If the Fréchet derivative $P'(x)$ is continuous in U and satisfies relation (3.9) with r defined by (3.7), then the sequence of iterates x_n defined by (3.10) lies in U_0 and converges to a solution x^* of equation $Px = 0$. The error estimate is given by relation (3.8).

Proof: Since $h = h(x) = -P'(x)^{-1}Px$ is evidently a uniform B-strategy for P, the proof follows immediately from Theorem 3.2. ∎

Remark 3.1 also applies to Theorem 2.3.

3.2 REMARK Condition (3.9) appears in a different form and with

$$B(s) = (2K)^{-1/2}s^{1/2}$$

in Kantorovič's hypothesis used in his convergence proof for Newton's method [see Altman (1983a), where K is the Lipschitz constant for $P'(x)$].

4. A GENERAL ITERATIVE METHOD OF CONTRACTOR DIRECTIONS

Let $P: D(P) \subset X \to Y$ be a nonlinear continuous mapping, and $\bar{q} < 1$ be positive. Suppose that for every $x \in U_0$, there exist an element $h = h(x)$ and an element $d(x,h)$ with the following properties.

$$(P(x_n + \varepsilon_n h(x_n)) - Px_n)/\varepsilon_n \to d(x_n, h(x_n)) \quad \text{as } \varepsilon_n \to 0+ \qquad (4.1)$$

whenever $\{x_n\}$ is a Cauchy sequence which lies in U_0 and $h(x_n)$ is bounded, and

$$\|d(x,h(x)) + Px\| \le \bar{q}\|Px\| \qquad (4.2)$$

for all $x \in U_0$;

$$\|h(x)\| \le B(\|Px\|) \qquad (4.3)$$

for some $B \in \mathbb{B}$ and all $x \in U_0$.

We can now define the following algorithm. Given $x_0 \in D(P)$, $\bar{q} < q < 1$ and $\bar{q}/q < \beta < 1$, suppose that $x_1, \ldots, x_n$ are already defined and $\Phi(1,h_n,x_n) \le q\|Px_n\|$, where $h_n = h(x_n)$ and

$$\Phi(\varepsilon,x,h) = \|P(x + \varepsilon h) - (1 - \varepsilon)Px\|/\varepsilon \tag{4.4}$$

Then we put $\varepsilon_n = 1$ and $x_{n+1} = x_n + \varepsilon_n h_n$. If $\Phi(1,x_n,h_n) > 1$, then there exists a positive $\varepsilon_n < 1$ such that

$$\beta q\|Px_n\| \le \Phi(\varepsilon_n,x_n,h_n) \le q\|Px_n\| \tag{4.5}$$

and we put

$$x_{n+1} = x_n + \varepsilon_n h_n \tag{4.6}$$

Then the sequence $\{x_n\}$ will be well defined for all n.

4.1 THEOREM Suppose that P is continuous and satisfies the hypotheses (4.1) to (4.3) for all $x \in U_0$ with radius r determined by relation (3.7). Then the sequence $\{x_n\}$ defined by (4.6) lies in U_0 and converges to a solution x* of equation $Px = 0$, and the error estimate (3.8) holds.

Proof: It is easy to see that the hypotheses of Lemma 2.1 are fulfilled. Hence, it follows that the sequence $\{x_n\}$ is a Cauchy sequence which lies in U_0 with radius r determined by relation (3.7). The estimate (3.8) results from (2.4). It remains to show that $Px_n \to 0$ as $n \to \infty$. To prove this, we suppose that

(a) The choice $\varepsilon_n = 1$ occurs an infinite number of n. Then obviously, $T = \infty$, and by virtue of Remark 7.1, the sequence $\{a_n\}$ is convergent to 0, and so is $\{\|Px_n\|\}$, by Lemma 2.1.

If (a) is not the case, then relation (4.5) applies for almost all n. Since

$$\Phi(\varepsilon_n,x_n,h_n) \le \varepsilon_n^{-1}\|P(x_n + \varepsilon_n h_n) - Px_n - \varepsilon_n d(x_n,h_n)\| + \bar{q}\|Px_n\| \tag{4.7}$$

by virtue of (4.2) with $x = x_n$ and $h_n = h(x_n)$, we get from (4.5) and (4.7),

$$(\beta q - \bar{q})\|Px_n\| \le \|P(x_n + \varepsilon_n h_n) - Px_n - \varepsilon_n d(x_n,h_n)\|/\varepsilon_n \tag{4.8}$$

Now suppose that $\varepsilon_n \to 0$ as $n \to \infty$. Since $\{x_n\}$ is a Cauchy sequence which lies in U_0 and $\{h_n\}$ is bounded, by virtue of (4.3) with $x = x_n$, it follows that $Px_n \to 0$ as $n \to \infty$, by virtue of (4.8) and (4.1). On the other hand, if $\{\varepsilon_n\}$ is not convergent to 0, then obviously $T = \infty$, and using the same reasoning as in case (a), we conclude that $Px_n \to 0$ as $n \to \infty$. This completes the proof. ∎

5. CONVERGENCE OF THE DAMPED NEWTON-KANTOROVIČ METHOD WITH APPROXIMATE SOLUTIONS OF THE LINEARIZED EQUATIONS FOR OPERATORS WITH CONTINUOUS FRÉCHET DERIVATIVES

The iterative method considered in Theorems 3.2 and 3.3 is based on the exact solution of the linearized equation, i.e., for each iteration step, we solve the linear equation

$$P'(x_n)h = -Px_n \qquad \text{for } n = 0, 1, \ldots$$

and h_n is supposed to be the exact solution of this equation.

However, in practice, it is usually not possible to solve the linear equation exactly, and one must solve it approximately. This means that the method which is in practical use and is supposed to be the Newton-Kantorovič method is in fact a different one.

Krasnosel'skii and Rutickii (1961) have proved the convergence of the Newton-Kantorovič method with approximate solutions of the linearized equations for operators with Hölder continuous Fréchet derivatives. The question remains open whether a proof can be given that does not utilize the Hölder continuity of the Fréchet derivative of the operator involved in the equation. However, the answer is positive for the damped method in question. To prove this result, we make use of the reasoning employed in Section 4.

We assume now that the Fréchet derivative $P'(x)$ of the nonlinear operator $P\colon D(P) \subset X \to Y$ exists and is continuous in $U = D(P) \cap \bar{S}(x_0, r)$. We also assume that there exists a positive $\bar{q} < 1$ with the following property. For each $x \in U_0$, there exists an element $h(x)$ such that

$$\|P'(x)h(x) + Px\| \le \bar{q}\|Px\| \tag{5.1}$$

and

$$\|h(x)\| \le B(\|Px\|) \tag{5.2}$$

for some $B \in \mathbb{B}$.

We can now define the algorithm. Given $x_0 \in D(P)$, $\bar{q} < q < 1$, and $\bar{q}/q < \beta < 1$, suppose that $x_1, \ldots, x_n$ are already defined. Then we put

$$x_{n+1} = x_n + \varepsilon_n h_n \tag{5.3}$$

where $h_n = h(x_n)$ is chosen so as to satisfy relations (5.1) and (5.2) with $x = x_n$, and ε_n is defined as follows. If $\Phi(1,x_n,h_n) \le q\|Px_n\|$, then we put $\varepsilon_n = 1$. If $\Phi(1,x_n,h_n) > q\|Px_n\|$, then there is a positive $\varepsilon_n < 1$ which satisfies relation (4.5), where $\Phi(\varepsilon,x_n,h_n)$ is defined by relation (4.4) with $x = x_n$ and $h = h_n$. The sequence $\{x_n\}$ will be well defined for all n, and the following theorem holds.

5.1 THEOREM Suppose that the Fréchet derivative $P'(x)$ exists and is continuous in U, and for each $x \in U_0$, there exists an element $h(x)$ satisfying relations (5.2) and (5.3). If the radius is determined by relation (3.7), then the sequence of iterates x_n defined by (5.3) lies in U_0 and converges to a solution x^* of equation $Px = 0$, and the error estimate (3.8) holds.

Proof: The proof is exactly the same as that of Theorem 4.1, where we put $d(x,h) = P'(x)h$. In particular, we get instead of relation (4.8) the following one.

$$(\beta q - \bar{q})\|Px_n\| \le \|P(x_n + \varepsilon_n h_n) - Px_n - \varepsilon_n P'(x_n)h_n\|/\varepsilon_n \tag{5.4}$$

Since

$$\|P(x_n + \varepsilon_n h_n) - Px_n - \varepsilon_n P'(x_n)h_n\|/\varepsilon_n$$

$$\le \|h_n\| \int_0^1 \|P'(x_n + \varepsilon_n t h_n) - P'(x_n)\|\, dt$$

it follows from the continuity of $P'(x)$ at $x = x^*$ (see Lemma 3.1) that the right-hand side of the latter inequality converges to 0 as $\varepsilon_n \to 0$. Hence, we obtain, by virtue of (5.4), that $\varepsilon_n \to 0$ implies $Px_n \to 0$ as $n \to \infty$, and we continue the proof in the same way as in Theorem 4.1. ∎

5.1 REMARK Krasnosel'skii and Rutickii (1961) have given a convergence proof for the method defined by (5.3) where $\varepsilon_n = 1$ for all n. However, their proof requires condition (5.1) to be replaced by the following one:

$$\|\bar{h}(x) - h\| \le \bar{q}\|P'(x)^{-1}Px\| \quad \text{with } 0 < \bar{q} < 1 \tag{5.5}$$

where $\bar{h}(x)$ is the approximate solution of the linear equation

$$P'(x)h + Px = 0 \quad \text{for } x = x_n \tag{5.6}$$

It is clear that condition (5.1), which yields the remainder for the approximate solution of equation (5.6), can easily be verified, while in practice it is rather more difficult to evaluate the error estimate (5.5).

Let us mention that Remark 3.1 also applies to Theorem 5.1.

6. AN ITERATIVE METHOD OF CONTRACTOR DIRECTIONS FOR NONDIFFERENTIABLE OPERATORS

The iterative methods discussed above require some kind of differentiability of the operator involved in the equation. However, it is also possible to define a convergent iterative method of contractor directions for nonlinear operators which are not differentiable but admit some kind of approximation by smooth operator. This kind of approximation is called Lipschitz approximation.

6.1 DEFINITION Let $P\colon D(P) \subset X \to Y$ be a nonlinear operator, then the operator $T\colon X \to Y$ is a Lipschitz approximation for P if the operator $Px - Tx$ is Lipschitz continuous; i.e., there exists a positive constant K such that

$$\|[Px - Tx] - [P\bar{x} - T\bar{x}]\| \le K\|x - \bar{x}\| \tag{6.1}$$

for all $x, \bar{x} \in D(P)$.

We assume that the continuous operator P has a Lipschitz approximation T which is differentiable in the sense of Fréchet and the Fréchet derivative $T'(x)$ is continuous in $U = D(P) \cap \bar{S}(x_0,r)$. We also assume that the Fréchet derivative $T'(x)$ has the following property. For each $x \in U_0 = D(P) \cap S(x_0,r)$, there exists an element $h(x) \in X$ such that

$$T'(x)h(x) + Px = 0 \tag{6.2}$$

and

$$\|h(x)\| \le B(\|Px\|) \tag{6.3}$$

where $B(s) = C \cdot s$, $C > 0$ being such that

$$KC < 1 \tag{6.4}$$

We can now define the following algorithm. Given $x_0 \in D(P)$, $KC/q < \beta < 1$, and $KC < q < 1$, suppose that $x_1, \ldots, x_n$ are already defined. Then we put $\varepsilon_n = 1$ if $\Phi(1,x_n,h_n) \le q\|Px_n\|$, where $\Phi(\varepsilon,x,h)$ is defined by formula (4.4). If $\Phi(1,x_n,h_n) > q\|Px_n\|$, then there exists a positive $\varepsilon_n < 1$ such that

$$\beta q\|Px_n\| \le \Phi(\varepsilon_n,x_n,h_n) \le q\|Px_n\| \tag{6.5}$$

In both cases, we put

$$x_{n+1} = x_n + \varepsilon_n h_n \tag{6.6}$$

Then the sequence $\{x_n\}$ will be well defined, and the following theorem holds.

6.1 THEOREM Suppose that the continuous operator P has a Lipschitz approximation T which satisfies relation (6.1) in U_0. Suppose also that the Fréchet derivative $T'(x)$ is continuous in U and satisfies the hypotheses (6.2) to (6.4) in U_0 with radius r determined by relation (3.7). Then the sequence of iterates x_n defined by (6.6) lies in U_0 and converges to a solution x^* of equation $Px = 0$, and the error estimate (3.8) holds.

Proof: It follows from relations (6.5) and (6.3) with $x = x_n$ that the hypotheses of Lemma 2.1 are fulfilled. Therefore, it remains to be shown that $Px_n \to 0$ as $n \to \infty$. We have

$$\|P(x_n + \varepsilon_n h_n) - (1 - \varepsilon_n)Px_n\|$$
$$\le \|T[P(x_n + \varepsilon_n h_n) - Tx_n - \varepsilon_n T'(x_n)h_n\|$$

$$+ \|[P(x_n + \varepsilon_n h_n) - T(x_n + \varepsilon_n h_n)] - [Px_n - Tx_n]\|$$

$$\le \|T(x_n + \varepsilon_n h_n) - Tx_n - \varepsilon_n T'(x_n)h_n\| + KC\varepsilon_n\|Px_n\|$$

by virtue of (6.1) and (6.3). Hence, we obtain (6.7),

$$(\beta q - KC)\|Px_n\| \le \|T(x_n + \varepsilon_n h_n) - Tx_n - \varepsilon_n T'(x_n)h_n\|/\varepsilon_n \tag{6.7}$$

by virtue of (6.5). Since the sequence $\{x_n\}$ converges to x^*, by Lemma 2.1, and $T'(x)$ is continuous at $x = x^* \in U$, it follows from (6.7) and from the boundedness of $\{h_n\}$ that $\varepsilon_n \to 0$ as $n \to \infty$ implies that $Px_n \to 0$ as $n \to \infty$, by using the same argument as in the proof of Lemma 3.1. We continue the proof of our theorem by repeating the same reasoning as in the proof of Theorem 3.1. ∎

Let us observe that the convergence proof of Theorem 6.1 is based on the fact that $h_n = h(x_n)$ is the exact solution of equation (6.2) with $x = x_n$. However, in practice, it is in general not possible to find the exact solution of the linear equation in question, and one must solve the equation approximately. With this in mind, we can define an iterative method of contractor directions which is based on the approximate solutions of equation (6.2) with $x = x_n$. For this purpose we assume that there exists a positive $\bar{q} < 1$ which has the following property. For each $x \in U_0$, there exists an element $h(x)$ such that

$$\|T'(x)h(x) + Px\| \le \bar{q}\|Px\| \tag{6.8}$$

and $h(x)$ satisfies relation (6.3). In addition, we assume that

$$KC + \bar{q} < 1 \tag{6.9}$$

where K is the constant in (6.1). Now we can define the following algorithm. Given $x_0 \in D(P)$ and

$$KC + \bar{q} < q < 1 \quad \text{and} \quad (KC + \bar{q})/q < \beta < 1 \tag{6.10}$$

supposing that $x_1, \ldots, x_n$ are already defined. Then we put $\varepsilon_n = 1$ or choose $\varepsilon_n < 1$ so as to satisfy relation (6.5) in the same way as in Theorem 6.1, provided that q and β are subject to relations (6.10). Then we define x_{n+1} by means of formula (6.6).

The following theorem holds for the method just defined.

6.2 THEOREM Suppose that the hypotheses of Theorem 6.1 are fulfilled with (6.2) replaced by (6.8), and q,β satisfying relations (6.10). If the radius r is determined by relation (3.7), then the sequence of iterates x_n lies in U_0 and converges to a solution x^* of equation $Px = 0$, and the error estimate (3.8) holds.

Proof: The following estimate results from (6.1), (6.3), and (6.8):

$$\|P(x_n + \varepsilon_n h_n) - (1 - \varepsilon_n)Px_n\| \le \|T(x_n + \varepsilon_n h_n) - Tx_n - \varepsilon_n T'(x_n)h_n\| + KC\varepsilon_n\|Px_n\| + \bar{q}\varepsilon_n\|Px_n\|$$

Hence, we obtain, by virtue of (6.5),

$$(\beta q - KC - \bar{q})\|Px_n\| \le \|T(x_n + \varepsilon_n h_n) - Tx_n - \varepsilon_n T'(x_n)h_n\|/\varepsilon_n \tag{6.11}$$

The proof is now exactly the same as that of Theorem 6.1, where relation (6.7) should be replaced by (6.11). ∎

7. THE B-PROPERTY

Let $P\colon\ D(P) \subset X \to Y$ be a nonlinear operator, where $D(P)$ is a vector space and X,Y are real or complex Banach spaces.

Denote by $\mathbb{B}$ the class of increasing continuous functions B such that $B(0) = 0$ and $B(s) > 0$ for $s > 0$; $\int_0^a s^{-1}B(s)\,ds < \infty$ for some positive a.

7.1 DEFINITION (Altman, 1977) Given $P\colon\ D(P) \subset X \to Y$, then $\Gamma_x(P) = \Gamma_x(P,q)$ is a set of contractor directions for P at $x \in D(P)$, which has the B-property, if for arbitrary $y \in \Gamma_s(P)$ there exist a positive number $\varepsilon = \varepsilon(x,y) \le 1$ and element $h = h(x,y) \in X$ such that

$$\|P(x + \varepsilon h) - Px - \varepsilon y\| \le q\varepsilon\|y\| \tag{7.1}$$

$$\|h\| \le B(\|y\|) \tag{7.2}$$

where $x + \varepsilon h \in D(P)$ and $q = q(P) < 1$ is some positive constant independent of $x \in D(P)$. The element h associated with $\Gamma_x(P)$ is called a strategic direction at $x \in D(P)$.

7.1 REMARK (Altman, 1984c) Let $\{a_n\}$ be a positive sequence defined as follows: $a_{n+1} = (1 - q\varepsilon_n)a_n$, for $n = 0, 1, \ldots$, where $0 < q < 1$ and $0 < \varepsilon \leq 1$.

Let B be some function from class $\mathbb{B}$. Then the series $\Sigma_{n=0}^{\infty} \varepsilon_n B(a_n)$ is convergent, and we have

$$\sum_{n=0}^{\infty} \varepsilon_n B(a_n) \leq q^{-1} \int_0^a s^{-1} B(s)\, ds \qquad \text{with } a = e^q a_0$$

and the remainder then satisfies the relation

$$\sum_{i=n}^{\infty} \varepsilon_i B(a_i) < q^{-1} \int_b^{b_n} s^{-1} B(s)\, ds$$

where $b_n = a_0 \exp(q(1 - t_n))$, $b = a_0 \exp(q(1 - T))$, $t_0 = 0$, $t_n = \Sigma_{i=0}^{n-1} \varepsilon_i$, and $T = \Sigma_{i=0}^{\infty} \varepsilon_i$.

Moreover, $a_n \to 0$ as $n \to \infty$ if and only if $T = \infty$.

APPENDIX 2

The Method of Contractor Directions for Nonlinear Elliptic Equations

1. A GENERAL EXISTENCE THEOREM

The purpose of this section is to present a general method of proving existence of solutions of nonlinear operator equations in Banach spaces. It is assumed that the corresponding linearized equations can be solved and the a priori estimates of their solutions are known. The general idea is adapted from Altman (1982b), where the application of the method of contractor directions is essential. The aim of this method is to provide a general framework for solving boundary value problems for nonlinear elliptic equations provided that we can solve the corresponding boundary value problems for the linearized equations for which the Schauder type estimates are known. As a model for applications, the Agmon et al. (1959, 1964) estimates for boundary value problems of linear elliptic partial differential equations are utilized.

Let $P\colon\ D(P) \subset X \to Y$ be a nonlinear operator, where $D(P)$ is a vector space, X,Y being real or complex Banach spaces. Let $S = S(x_0,r)$ be an open ball with center $x_0 \in D(P)$ and radius $r > 0$, and put $U_0 = U \cap S$, where U is a vector space with $U \subset D(P)$.

We make the following assumptions.

(A_1) There exists a family $\{\tilde{P}\}$ of mappings $\tilde{P}\colon\ U_0 \to Y$ which has the following properties. For arbitrary $\eta < 0$, there exists $\tilde{P} \in \{\tilde{P}\}$ such that

$$\|Px - \tilde{P}x\| \leq \eta \qquad \text{for all } x \in U_0 \tag{1.1}$$

For each $\tilde{P} \in \{\tilde{P}\}$ there exists a mapping

$$\tilde{P}': \quad U_0 \to B(X,Y)$$

the space of bounded linear operators from X into Y, such that

$$\|\tilde{P}(x + \varepsilon h) - \tilde{P}x - \varepsilon\tilde{P}'(x)h\| \leq K\varepsilon^2\|h\|^2 \tag{1.2}$$

for all x, $x + \varepsilon h \in U_0$, $h \in U$, where K is a global constant independent of $\tilde{P}'$, and $0 < \varepsilon \leq 1$.

(A_2) There exists a family {A} of mappings A: $U_0 \to B(X,Y)$ which has the following property: For arbitrary $\delta > 0$ and arbitrary $\tilde{P}'$, there exists $A \in \{A\}$ such that A: $U_0 \to B(X,Y)$ and

$$\|[\tilde{P}'(x) - A(x)]h\| \leq \delta\|h\| \tag{1.3}$$

for all $x \in U_0$ and $h \in X$.

(A_3) Let $x \in U_0$ be arbitrary. Then we assume that there exists a sequence $\{A_m(x)\}$ of linear continuous mappings from X into Y such that

$$\lim\|A(x) - A_m(x)\|_{X,Y} = 0 \tag{1.4}$$

(A_4) There exists a vector space $V \subset Y$ which has the following property: For arbitrary $x \in U_0$ and $\omega > 0$, there exists $y_\omega \in V$ such that

$$\|Px - y_\omega\| \leq \omega \tag{1.5}$$

(A_5) For arbitrary $x \in X_0$ and y satisfying (A_4) the equation

$$A_m(x)h + y = 0 \tag{1.6}$$

has a solution $h \in U$ such that

$$\|h\| \leq C\|y\| \tag{1.7}$$

where C is a global constant independent of both x and m.

For $x \in U_0$, let $h \in U$ be a solution of equation (1.6) with y satisfying (1.5) and m, ω to be determined. Let $C_1 > C$; then we obtain, by virtue of (1.7),

$$\|h\| \le C\|y\| \le C\|Px\| + C\|Px - y\|$$
$$\le C_1\|Px\|$$

Hence, we get for sufficiently small ω,

$$\|h\| \le C_1\|Px\| \tag{1.8}$$

if $\|Px - y\| \le (C_1 - C)\|Px\|/C$.

1.1 LEMMA Let q be an arbitrary number with $0 < q < 1$. Then $0 < \varepsilon \le 1$ can be chosen so as to satisfy

$$\|P(x + \varepsilon h) - (1 - \varepsilon)Px\| \le q\varepsilon\|Px\| \tag{1.9}$$

$$\|h\| \le C_1\|Px\| \tag{1.10}$$

$$\bar{x} \in U_0 \tag{1.11}$$

where $\bar{x} = x + \varepsilon h$ and $x \in U_0$.

Proof: $P(x + \varepsilon h) - Px + \varepsilon Px$

$$= P(x + \varepsilon h) - \tilde{P}(x + \varepsilon h)] - (1 - \varepsilon)(Px - \tilde{P}x)$$
$$+ [\tilde{P}(x + \varepsilon h) - \tilde{P}x - \varepsilon\tilde{P}'(x)h]$$
$$+ \varepsilon[\tilde{P}'(x) - A(x)]h + \varepsilon[A(x) - A_m(x)]h$$
$$+ \varepsilon[A_m(x)h + y] + \varepsilon[Px - y]$$
$$+ \varepsilon[\tilde{P}x - Px]$$

By using (1.1) with $2\eta \le \varepsilon q\|Px\|/5$ we get an estimate for the first two terms of the above identity and the last term. We obtain from (1.2) that

$$\|\tilde{P}(x + \varepsilon h) - \tilde{P}x - \varepsilon\tilde{P}'(x)h\| \le \varepsilon q\|Px\|/5 \tag{1.12}$$

if

$$\varepsilon \le q\|Px\|/\|h\|^2 5K \tag{1.13}$$

Relation (1.3) yields

$$\|[\tilde{P}'(x) - A(x)]h\| \le q\|Px\|/5$$

if $\delta \le q/5C_1$, by virtue of (1.10), and

$$\|[A(x) - A_m(x)]h\| \le q\|Px\|/5$$

if m is sufficiently large, by virtue of (1.4) and (1.10). Finally, we get the following estimate for the seventh term:

$$\|Px - y\| \le q\|Px\|/5$$

if $\omega \le q\|Px\|/5$, by virtue of (1.5). By combining the above estimates we obtain relation (1.9). Thus, relation (1.9) holds with $0 < \varepsilon \le 1$ satisfying (1.13). Condition (1.11) is satisfied if $0 < \varepsilon \le 1$ is such that

$$\varepsilon C_1\|Px\| < r - \|x - x_0\| \tag{1.14}$$

by virtue of (1.10) which follows from (1.8). Thus, $0 < \varepsilon \le 1$ can be chosen so as to satisfy relations (1.13) and (1.14). ■

1.2 LEMMA Let the sequence $\{x_n\} \subset U_0$ be such that

$$\|x_n - x\| \to 0 \qquad \|Px_n - Px\| \to 0 \qquad \|x_n - x\| < r \tag{1.15}$$

for some x with $\|Px\| \neq 0$.

Let $0 < q' < q < 1$ and $C < C_1 < \bar{C}_1 < \bar{C}$. Then there exist $0 < \varepsilon \le 1$ and $\bar{x} \in U_0$ such that

$$\|P\bar{x} - (1 - \varepsilon)Px\| \le q'\varepsilon\|Px\| < q\varepsilon\|Px\| \tag{1.16}$$

and

$$\|\bar{x} - x\| \le \varepsilon\bar{C}_1\|Px\| < \varepsilon\bar{C}\|Px\| \tag{1.17}$$

Proof: Choose $0 < \varepsilon \le 1$ such that

$$\varepsilon C_1\|Px\| < r - 1\|x - x_0\| \tag{1.18}$$

and n so large that

$$\|x_n - x\| < r - \|x - x_0\| - \varepsilon C_1\|Px\| \tag{1.19}$$

Having chosen $0 < \varepsilon \le 1$, we then move with x_n, Px_n closer to x, Px, respectively, so as to satisfy the following relations.

$$\|Px_n - Px\| \le \varepsilon q'\|Px\|/2 \tag{1.20}$$

and

$$\|x_n - x\| \le \varepsilon(\bar{C}_1 - C_1)\|Px\| \tag{1.21}$$

Let $y_n \in V$ be such that

$$\|Px_n - y_n\| \le \omega \tag{1.22}$$

with ω to be determined. Now let h_n be a solution of the equation

$$A_m(x_n)h + y_n = 0 \tag{1.23}$$

with $m = m(n)$ to be determined. Then we get, by virtue of (1.7),

$$\begin{aligned}\|h_n\| &\le C\|y_n\| \\ &\le C\|Px\| + C\|Px_n - Px\| + C\|Px_n - y_n\|\end{aligned}$$

Hence, we obtain for $C_1 > C$,

$$\|h_n\| \le C_1\|Px\| \tag{1.24}$$

if ω in (1.22) is sufficiently small and m,n are sufficiently large with n satisfying relations (1.19) to (1.21).

In order to prove relation (1.16), the following identity is used.

$$\begin{aligned}&P(x_n + \varepsilon h) - Px_n + \varepsilon Px_n \\ &= [P(x_n + \varepsilon h_n) - \tilde{P}(x_n + \varepsilon h_n)] - (1 - \varepsilon)(Px - \tilde{P}x) \\ &\quad + [\tilde{P}(x_n + \varepsilon h_n) - \tilde{P}x_n - \varepsilon\tilde{P}'(x_n)h_n] \\ &\quad + \varepsilon[\tilde{P}'(x_n) - A(x_n)]h_n + \varepsilon[A(x_n) - A_m(x_n)]h_n \\ &\quad + \varepsilon[A_m(x_n)h_n + y_n] + \varepsilon[Px_n - y_n] \\ &\quad + \varepsilon[\tilde{P}x - Px]\end{aligned}$$

From (1.1) with $x = x_n$ and $2\eta \le \varepsilon q'\|Px_n\|/10$, we get an estimate for the first two terms of the above identity and the last term. We obtain from (1.2) with $x = x_n$, $h = h_n$, and $q = q'$ that

$$\|\tilde{P}(x_n + \varepsilon h_n) - \tilde{P}x_n - \varepsilon\tilde{P}'(x_n)h_n\| \le \varepsilon q'\|Px\|/10 \tag{1.25}$$

if, by virtue of (1.24),

$$\varepsilon \le (q'/10KC_1^2)\|Px\| \tag{1.26}$$

Relation (1.3) with $x = x_n$ and $h = h_n$ yields

$$\|[\tilde{P}'(x_n) - A(x_n)]h_n\| \le q'\|Px\|/10$$

if $\delta \le q'/10C_1$, by virtue of (1.24), and

$$\|[A(x_n) - A_m(x_n)]h_n\| < q'\|Px\|/10$$

if $m = m(n)$ is sufficiently large, by virtue of (1.4) with $x = x_n$ and (1.24). Finally, we get the following estimate for the seventh term:

$$\|Px_n - y_n\| \le q'\|Px\|/10$$

if $\omega \le q'\|Px\|/10$, by virtue of (1.5) with $x = x_n$ and $y_\omega = y_n$. By combining the above estimates, we obtain that

$$\|P(x_n + \varepsilon h_n) - (1 - \varepsilon)Px_n\| \le q'\varepsilon\|Px\|/2 \tag{1.27}$$

It follows from (1.27) and (1.20) that relation (1.16) holds with $\bar{x} = x_n + \varepsilon h_n$ if n is sufficiently large. Since

$$\begin{aligned}\|\bar{x} - x\| &= \|x_n + \varepsilon h_n - x\| \\ &\le \|x_n - x\| + \varepsilon\|h_n\| \le \varepsilon\bar{C}_1\|Px\|\end{aligned}$$

by virtue of (1.21) and (1.24), relation (1.17) is satisfied. Finally, it is clear that $\bar{x} \in U_0$, by virtue of (1.19) and (1.24), and the proof is completed. ■

1.1 THEOREM In addition to the hypotheses (A_1) to (A_5), suppose that the mapping P is closed on U_1, the closure of $U_0 = U \cap S(x_0,r)$, and the radius r is such that

$$r > \bar{r} = (1 - q)^{-1}\bar{C}\|Px_0\| \exp(1 - q) \qquad C < \bar{C}$$

where $\bar{C}$ and $0 < q < 1$ are arbitrary. Then there exists a solution $x \in U_1$ of the equation $Px = 0$.

Proof: The proof is similar to that in Altman (1984a). We construct a sequence $\{x_n\} \subset U_0$ and a sequence $\{\varepsilon_n\}$ with $0 < \varepsilon_n \le 1$ such that

$$\|Px_{n+1} - (1 - \varepsilon_n)Px_n\| \le q\varepsilon_n\|Px_n\| \tag{1.28}$$

$$\|x_{n+1} - x_n\| \le \varepsilon_n\bar{C}\|Px_n\| \tag{1.29}$$

as follows. Suppose that $x_0, x_1, \ldots, x_n$ and $\varepsilon_0, \varepsilon_1, \ldots, \varepsilon_{n-1}$ are known. Then it follows from Lemma 1.1 with $x = x_n$ that there exist $0 < \bar{\varepsilon} \le 1$ and $\bar{x}_{n+1} = x_n + \bar{\varepsilon}_n h_n$ such that

$$\bar{x}_{n+1} \in U_0 \tag{1.30}$$

$$\|P\bar{x}_{n+1} - (1 - \bar{\varepsilon}_n)Px_n\| \tag{1.31}$$

$$\|\bar{x}_{n+1} - x_n\| \le \bar{\varepsilon}_n\bar{C}\|Px_n\| \tag{1.32}$$

Then we choose ε_n to be approximately (up to a constant factor $0 < c < 1$) the largest number $\bar{\varepsilon}_n$ for which x_{n+1} exists and satisfies relations (1.30) to (1.32). Now it follows from Lemma 2.1, Appendix 1, that $\|x_n\| < \bar{r} < r$ for all n, and there exist x and z such that

$$x_n \to x \qquad \text{and} \qquad Px_n \to z \qquad \text{as } n \to \infty$$

Since P is closed on U_1, it follows that $z = Px$. It also follows from Lemma 2.1, Appendix 1, that if $\{\varepsilon_n\}$ is not convergent to 0, then $Px = 0$. Therefore, suppose that $Px \neq 0$ and

$$\varepsilon_n \to 0 \qquad \text{as } n \to \infty \tag{1.33}$$

It results from Lemma 1.2 that there exist $0 < \varepsilon \le 1$ and $\bar{x} \in U_0$ which satisfy relations (1.16) and (1.17). Since

$$\|x_n - x\| \to 0 \qquad \text{and} \qquad \|Px_n - Px\| \to 0 \qquad \text{as } n \to \infty$$

and the sequence $\{\|Px_n\|\}$ is decreasing, it follows that

$$\|P\bar{x} - (1 - \varepsilon)Px_n\| \le q\varepsilon\|Px_n\|$$

and

$$\|\bar{x} - x_n\| \leq \varepsilon \bar{C} \|Px_n\|$$

for all sufficiently large n. But the approximate maximality choice of ε_n implies that $\varepsilon_n \geq c\varepsilon$, where $0 < c < 1$. But this is a contradiction to (1.33) and the proof of the theorem is completed. ∎

The following remark can be useful in practical applications of Theorem 1.1.

1.1 REMARK By using Lemma 1.1 with $x = x_n$, we can construct a sequence $\{x_n\} \subset U_0$ such that

$$x_{n+1} = x_n + \varepsilon_n h_n \qquad 0 < \varepsilon_n \leq 1$$

$$\|P(x_n + \varepsilon_n h_n) - (1 - \varepsilon_n)Px_n\| \leq q\varepsilon_n \|Px_n\|$$

$$\|h_n\| \leq \bar{C} \|Px_n\|$$

for all n. If $\Sigma_{n=0}^{\infty} \varepsilon_n = \infty$, then $x = x_0 + \Sigma_{n=0}^{\infty} \varepsilon_n h_n$ is a solution of the equation $Px = 0$, and $x \in U_1$.

Proof: The proof results from Lemma 2.1. ∎

2. THE AGMON-DOUGLIS-NIRENBERG ESTIMATES

In order to illustrate how to apply Theorem 1.1 to nonlinear boundary value problems of elliptic equations, the estimates obtained by Agmon et al. (1959; 1964) are needed.

Let Ω be a bounded domain in the N-dimensional euclidean space E_N and denote by $\partial\Omega$ its boundary and by $\bar{\Omega}$ its closure. We use the notation $x = (x_1, \ldots, x_N) \in E_N$.

$$D_i = \frac{\partial}{\partial x_i} \qquad D = (D_1, \ldots, D_N)$$

$$D^\beta = D_1^{\beta_1} \cdots D_N^{\beta_N} \qquad |\beta| = \Sigma \beta_j$$

We introduce the following Hölder norms for $f \in C^\ell(\bar{\Omega})$, the space of functions possessing continuous derivatives up to order $\ell \geq 0$.

$$|f|_\ell = |f|_\ell^\Omega = \sum_{j=0}^{\ell} \text{lub } |D^j f|^\Omega$$

$$[f]_{\ell+2} = \text{lub } \frac{|D^\ell f(x) - D^\ell f(x')|}{|x - x'|^\alpha} \qquad 0 < \alpha < 1$$

where the lub is over $x \neq x'$ in Ω and all derivatives of order ℓ, and the norm

$$|f|_{\ell+\alpha} = |f|_\ell + [f]_{\ell+\alpha}$$

Consider the boundary value problem

$$L(x)u(x) = f(x) \qquad x \in \Omega \tag{2.1}$$

$$B_j(x)u(x) = g_j(x) \qquad x \in \partial\Omega \qquad j = 1, \ldots, m \tag{2.2}$$

where

$$L = \sum_{|\beta| \leq 2m} a_\beta(\cdot)D^\beta$$

is a uniformly elliptic operator satisfying the conditions of Agmon et al. (1959, 1964), and

$$B_j = \sum_{|\gamma| \leq m_j} b_{j,\gamma}(x)D^\gamma \qquad j = 1, \ldots, m$$

satisfy the complementing condition. We also assume that, for j = 1, ..., m,

$$|u_n - u|_{\ell+\alpha} \to 0 \tag{2.3}$$

implies $|B_j(x)(u_n(x) - u(x))|_0^{\partial\Omega} \to 0$, $x \in \partial\Omega$. We assume that f and the coefficients of L have $|\cdot|_{\ell-2m+\alpha}$ norms bounded, while g_j and the coefficients of B_j have their $|\cdot|_{\ell-m_j+\alpha}$ bounded. Suppose that problems (2.1) and (2.3) have a unique solution u, such that

$$|u|_{\ell+\alpha}^\Omega \leq C(|f|_{\ell-2m+\alpha}^\Omega + \Sigma|g_j|_{j-m_j+\alpha}^{\partial\Omega})$$

where C is some constant (see Agmon et al., 1959; 1964).

3. THE BOUNDARY VALUE PROBLEM FOR NONLINEAR ELLIPTIC EQUATIONS

Consider a nonlinear differential equation of order 2m in N-space,

$$Pu \equiv F(x,u,Du,\ldots,D^{2m}u) = 0 \qquad x \in \Omega \tag{3.1}$$

satisfying m linear boundary conditions

$$B_j(x)u(x) = g_j(x) \qquad \text{for } j = 1, \ldots, m \qquad x \in \partial\Omega \tag{3.2}$$

as in (2.2). We assume that the first variation of the operator F is of the form

$$P'(u)h \qquad L(u)h(x) = \sum_\beta \frac{\partial F}{\partial D^\beta u}(x,u(x),\ldots,D^{2m}U(x))D^\beta h(x) \tag{3.3}$$

which is a linear (in h) elliptic operator. Let X be the Banach space of functions with Hölder norm $|\cdot|_{\ell+\alpha}$, defined in Section 2, and let Y be the Banach space of functions with Hölder norm $|\cdot|_{\ell-2m+\alpha}$.

Let $U \subset X$ be a vector space of sufficiently smooth functions. Put $U_0 = U \cap S(u_0,r)$, where $u_0 \in U$, and assume that U is contained in the domain of P given by (3.1). Denote by U_1 the closure of U_0. Consider the operator P defined as in (3.1) and assume that P maps U_1 into Y. For the sake of simplicity consider an application of Theorem 1.1 provided that $\tilde{P} = P$ and $A_m = \tilde{F}' = F'$; i.e., no approximations occur there. Thus conditions (1.1), (1.3), and (1.4) are deleted. Therefore, equations (1.6) should be replaced by the following:

$$P'(u)h = y \qquad \text{and} \qquad B_j(x)h(x) = 0 \qquad x \in \partial\Omega \tag{3.4}$$
$$j = 1, \ldots, m$$

with $h \in U$, and $y \in V$ approximating Pu, where V is a smooth vector space contained in Y as in condition (1.5), with x replaced by u. We have to assume that the Agmon et al. (1959, 1964) estimates (2.4) of Section 2 hold for equation (3.4) with a constant C independent of $u \in U_0$. Condition (1.2) should be replaced by the following:

$$\|P(u + \varepsilon h) - Pu - \varepsilon P'(u)h\| \leq K\varepsilon^2\|h\|^2$$

for all u, $u + h \in U_0$, $h \in U$, $0 < \varepsilon \leq 1$.

We also assume that

$$B_j(x)u_0(x) = g_j(x)$$

for $x \in \partial\Omega$ and $j = 1, \ldots, m$, so that the sequence $u_0, u_1, \ldots, u_n,$ ... constructed in the proof of Theorem 1.1 will satisfy the boundary condition

$$B_j(x)u_n(x) = g_j(x) \qquad \text{for } x \in \partial\Omega$$

for all n, by virtue of (3.4). Hence, by virtue of (2.3), we obtain for the limit u of $\{u_n\}$ that

$$B_j(x)u(x) = g_j(x) \qquad x \in \partial\Omega \qquad j = 1, \ldots, m$$

Notice that the boundary condition of (3.4) imposed on h implies that condition (1.7) is satisfied, by virtue of (2.4) with f replaced by y and u by h. Thus, relation (2.4) yields

$$\|h\|^{\Omega}_{\ell+\alpha} \leq C\|y\|_{\ell-2m+\alpha}$$

Let us mention that similar results can be obtained by utilizing the Agmon et al. (1959; 1964) estimates for Sobolev norms.

APPENDIX 3

An Application of the Method of Contractor Directions to the Boltzmann Equation for Neutron Transport in L^1 Space

1. INTRODUCTION

The Boltzmann equation provides a basic mathematical tool in many areas of physics, in particular, in kinetic theory of gases, statistical mechanics, neutron transport theory, etc. There is an extensive literature devoted to the linear Boltzmann equation (see Case and Zweifel, 1967) but considerably less is known about the nonlinear case. Our main focus is on the nonlinear Boltzmann equation for the neutron transport problem in a bounded space domain. More precisely, we consider the following nonlinear integro-differential equation,

$$\frac{\partial N}{\partial t} + v \cdot \operatorname{grad}_x N + \sigma(t,x,v,N) + \int_{\Omega_2} \sigma_s(t,x,v,v',N(t,x,v'))dv' + q(t,x,v) \tag{1.1}$$

$t > 0$, $x \in \Omega_1$, $v \in \Omega_2$, where Ω_1 and Ω_2 are bounded domains in the m-dimensional euclidean space, R^m $(m = 1, 2, \ldots)$, $N \equiv N(t,x,v)$ is the neutron density function, t is time, x is position, v is velocity, $q(t,x,v)$ is the source function, and σ and σ_s, being in general nonlinear functions of N, correspond to the total and scattering cross sections, respectively. Equation (1.1) is considered under the following boundary conditions:

$$N(t,x,v) = 0 \quad \text{for } x \text{ on } \partial\Omega_1 \tag{1.2}$$

$t > 0$ and incoming v,

$$N(0,x,v) = 0(x,v) \quad \text{for } x \text{ in } \Omega_1,\ v \text{ in } \Omega_2 \tag{1.3}$$

If $\sigma(t,x,v,N) = \sigma_0(t,x,v)N$ and $\sigma_s(t,x,v,v',N) = \sigma_{0s}(t,x,v,v')N$, then equation (1.1) is the well-known linearized Boltzmann equation for neutron transport in a bounded space domain.

Pao (1973) investigated (1.1) to (1.3) in L^1 and proved existence and uniqueness of both global and local solutions. He combined the concept of semi-inner product and the contraction principle. To this end he assumed the nonlinear functions σ and σ_s to be Lipschitz continuous in N. We also use the same semi-inner product argument, but instead of the contraction principle we apply our method of contractor directions (Altman, 1980) and some ideas from the theory of quasilinear and nonlinear evolution equations in nonreflexive Banach spaces (see Altman, 1982b). The advantage of our method is that the class of nonlinear Lipschitz continuous mappings in the nonlinear part of the equation (1.1) can be replaced by locally Lipschitz ones. This seems to be justified from the physical point of view because, for the sake of mathematical convenience, the functions that occur in physics are usually assumed to be smooth.

2. A GENERALIZATION OF THE BOLTZMANN EQUATION

Because of various modifications of the nonlinear part in equation (1.1), which occur in handling different problems in physics, it makes sense to introduce the following abstract version of the Boltzmann equation in a Banach function space X.

$$PU \equiv \frac{\partial U}{\partial t} + v \cdot \operatorname{grad}_x U + FU = 0 \tag{2.1}$$

where $U = U(t,x,v)$ and $F\colon\ D(F) \subset X \to X$ is a nonlinear operator, where $t > 0$, $x \in \Omega$, $v \in \Omega_2$ (bounded domains in m-dimensional euclidean space). It is convenient to choose $X = L^1(G)$, the space of Lebesgue integrable functions defined on $G = (0,T) \times \Omega$, where

$\Omega = \Omega_1 \times \Omega_2$. The norm in $L^1(G)$ is defined by $\|u\| = \int_G |u(z)|dz$, $u \in L^1(G)$. Assume that $\Phi(x,v) \in C^1(\bar{\Omega})$ satisfies the boundary condition (1.2), where $C^1(\bar{\Omega})$ is the class of continuously differentiable functions. Introducing a new variable $U = e^{-\lambda t}(N - \Phi)$ for some real number λ to be determined, we transform problems (1.1) to (1.3) into the following (see Pao, 1973):

$$U_t + v \cdot \text{grad}_x U + \lambda U = f(t,x,v,U) \tag{2.2}$$

$$U(t,x,v) = 0 \tag{2.3}$$

for $x \in \partial\Omega_1$, $t > 0$, and incoming velocity v,

$$U(0,x,v) = 0 \qquad \text{for } (x,v) \in \Omega \tag{2.4}$$

where $U_t = \partial U/\partial t$, $(t,x,v) \in G$, and

$$f(t,x,v,U) = f_1(t,x,v,U) + f_2(t,x,v,U) - e^{-\lambda t}v \cdot \text{grad}_x \Phi$$

$$f_1(t,x,v,U) = -e^{-\lambda t}\sigma(t,x,v,e^{\lambda}U + \Phi)$$

$$f_2(t,x,v,U) = e^{-\lambda t} \int_{\Omega_2} \sigma_s(t,x,v,v',e^{\lambda t}U(t,x,v') + \Phi(x,v'))dv' + e^{-\lambda t}q(t,x,v)$$

Consider the linear operator A defined in $L^1(\bar{G})$ by

$$AU = U_t + v \cdot \text{grad}_x U + \lambda U \tag{2.5}$$

with domain $D(A) = \{U \in C^1(\bar{G})\}$, where U satisfies (2.3) and (2.4).

Thus equation (2.1) can be written as follows:

$$PU = AU + FU = 0 \tag{2.6}$$

where A is given by (2.5). We need the following lemmas proved by Pao (1973) by means of the concept of semi-inner product.

2.1 LEMMA (Pao, 1973) The following inequality holds for A given by (2.5).

$$\lambda\|U\| \le \|AU\| \qquad \text{for } U \in D(A) \tag{2.7}$$

2.2 LEMMA (Pao, 1973) Given any $g(t,x,v)$ in $L^1(G)$ with $g(t,x,v) = 0$ if x is outside of Ω_1, problem

$$U_t + v \cdot \operatorname{grad}_x U + \lambda U = g(t,x,v) \tag{2.8}$$

along with (2.3) and (2.4), where λ is an arbitrary real number, has a solution $U(t,x,v)$ given by

$$U(t,x,v) = \int_0^t e^{-\lambda(t-\tau)} g(\tau,\ x - v(t - \tau),\ v)\, d\tau \tag{2.9}$$

Proof: Let $g \in D(A)$ be given. By introducing new variables $x' = x - vt$, $\tau = t$, we obtain from (2.8),

$$U_\tau(\tau,\ x' + v\tau,\ v) + \lambda U(\tau,\ x' + v\tau,\ v) = g(\tau,\ x' + v\tau,\ v)$$

Hence, multiplying by $e^{\lambda\tau}$ and integrating from 0 to t, we get

$$e^{\lambda\tau} U(t,\ x' + vt,\ v) - U(0,x',v) = \int_0^t e^{\lambda\tau} g(\tau,\ x' + v\tau,\ v)\, d\tau$$

Let $x' = x - vt$. Then (2.9) follows from initial condition (2.4). We see that $U(t,x,v)$ given by (2.9) is in $C^1(\bar{G})$. Since $g(t,x,v) = 0$ if $x \notin \Omega_1$, it follows that $g(t,\ x - v(t - \tau),\ v) = 0$ if x is not on $\partial\Omega_1$ and v is incoming. Thus $U(t,x,v)$ satisfies the boundary condition (2.3). Hence, $U(t,x,v) \in D(A)$ is a solution to problems (2.8), (2.3), and (2.4). The statement of Lemma 2.2 holds for arbitrary g in $L^1(G)$, since $D(A)$ is dense in $L^1(G)$. Then $U \in D(A_c)$, A_c being the closure of A. But for our purpose it suffices to consider only $g \in D(A)$. ∎

Let $S = S(u_0,r)$ be an open ball with center $u_0 \in D(A)$ and radius $r > 0$, and put $V_0 = D(A)S$.

We make the following assumptions.

(A_1) There exists a family $\{\tilde{F}\}$ of mappings $\tilde{F}$: $V_0 \to L^1(G)$ which has the following property: For arbitrary F and $\eta > 0$, there exists $\tilde{F} \in \{\tilde{F}\}$ such that

$$\|Fu - \tilde{F}u\| \le \eta \tag{2.10}$$

for all $u \in V_0$.

(A_2) Each $\tilde{F} \in \{\tilde{F}\}$ is locally Lipschitz continuous with Lipschitz constant K independent of both $\tilde{F} \in \{\tilde{F}\}$ and $u \in V_0$; i.e., there exists $0 < \rho \leq 0$ such that

$$\|\tilde{F}(u + \varepsilon h) - \tilde{F}u\| \leq K\varepsilon\|h\| \tag{2.11}$$

for all $u \in V_0$, $h \in D(A)$ and all $0 < \varepsilon \leq \rho \leq 1$ such that $u + \varepsilon h \in V_0$.

Let $y \in D(A)$ be given. Then equation

$$Ah + y = 0 \tag{2.12}$$

has a solution $h \in D(A)$ such that

$$\|h\| \leq \lambda^{-1}\|y\| \tag{2.13}$$

by virtue of Lemmas (2.1) and (2.2). Then we obtain for $C > \lambda^{-1}$,

$$\|h\| \leq \lambda^{-1}\|y\| \leq \lambda^{-1}\|Pu\| + (C - \lambda^{-1})\|Pu - y\| \leq C\|Pu\|$$

if $\|Pu - y\| \leq \|Pu\|/(C - \lambda^{-1})$. Hence, we get

$$\|h\| \leq C\|Pu\| \tag{2.14}$$

2.3 LEMMA Let $u \in V_0$ be given and let q be an arbitrary number with $0 < q < 1$. Then $0 < \varepsilon \leq \rho \leq 1$ can be chosen so as to satisfy

$$\|P(u + \varepsilon h) - (1 - \varepsilon)Pu\| \leq q\varepsilon\|Pu\| \tag{2.15}$$

$$\|h\| \leq C\|Pu\| \tag{2.16}$$

$$\bar{u} \in V_0 \tag{2.17}$$

where $\bar{u} = u + \varepsilon h$.

Proof: Put $\tilde{P}u = Au + \tilde{F}u$. Then

$$\begin{aligned} P(u + \varepsilon h) - Pu + \varepsilon Pu = {} & \{P(u + \varepsilon h) - \tilde{P}(u + \varepsilon h)\} \\ & - (1 - \varepsilon)(Pu - \tilde{P}u) \\ & + [\tilde{F}(u + \varepsilon h) - \tilde{F}u] + \varepsilon(Ah + y) \\ & + \varepsilon(Pu - y) + \varepsilon(\tilde{P}u - Pu) \end{aligned}$$

Taking into account that $Pu - \tilde{P}u = Fu - \tilde{F}u$, we obtain the following estimate:

$$\|P(u + \varepsilon h) - (1 - \varepsilon)Pu\| \le 2\eta + K\varepsilon\|h\| + \eta$$
$$\le \varepsilon K\lambda^{-1}\|y\| + 3\eta$$
$$\le \varepsilon K\lambda^{-1}\|Pu\| + \varepsilon K\lambda^{-1}\eta + 3\eta$$

Hence, we obtain relation (2.15) if $K\lambda^{-1} < q$ and $\varepsilon K\lambda^{-1}\eta + 3\eta \le \varepsilon(q - K\lambda^{-1})\|Pu\|$, where $\|Pu - y\| \le \eta$ and $\|Fu - \tilde{F}u\| \le \eta$, by virtue of (2.10), (2.11). Relation (2.16) follows from (2.14), where h is a solution of equation (2.12). For relation (2.17) to be satisfied we choose $0 < \varepsilon \le \rho \le 1$ such that

$$\varepsilon C\|Pu\| < r - \|u - u_0\| \tag{2.18}$$

Then we get

$$\|u + \varepsilon h - u_0\| \le \|u - u_0\| + \varepsilon C\|Pu\| < r$$

by virtue of (2.14), and the proof of the lemma is completed. ∎

2.4 LEMMA Let the sequence $\{u_n\} \subset V_0$ be such that

$$\|u_n - u\| \to 0 \qquad \|Pu_n - Pu\| \to 0 \qquad \|u_n - u\| < r \tag{2.19}$$

for some u with $\|Pu\| \ne 0$.

Let $0 < q' < q < 1$ and $C < C_1$. Then there exist $0 < \varepsilon \le 1$ and $\bar{u} \in V_0$ such that

$$\|P\bar{u} - (1 - \varepsilon)Pu\| \le q'\varepsilon\|Pu\| < q\varepsilon\|Pu\| \tag{2.20}$$

and with $\bar{C} > C_1$,

$$\|\bar{u} - u\| \le \varepsilon C_1\|Pu\| < \varepsilon\bar{C}\|Pu\| \tag{2.21}$$

Proof: Choose $0 < \varepsilon \le 1$ such that

$$\varepsilon C_1\|Pu\| < r - \|u - u_0\| \tag{2.22}$$

and n so large that

$$\|u_n - u\| < r - \|u - u_0\| - \varepsilon C_1\|Pu\| \tag{2.23}$$

Having chosen $0 < \varepsilon \le 1$, we then move with u_n, Pu_n closer to u, Pu, respectively, so as to satisfy the following relations,

$$\|Pu_n - Pu\| \le \varepsilon q'\|Pu\|/2 \tag{2.24}$$

and

$$\|u_n - u\| \le \varepsilon(\bar{C}_1 - C_1)\|Pu\| \tag{2.25}$$

Let $y_n \in D(A)$ be such that

$$\|Pu_n - y_n\| \le \eta_n \tag{2.26}$$

with η_n to be determined. Now let h_n be a solution of the equation

$$Ah + y_n = 0 \tag{2.27}$$

Then we obtain, by virtue of (2.14) with $u = u_n$,

$$\begin{aligned}\|h_n\| &\le C\|y_n\| \\ &\le C\|Pu\| + C\|Pu_n - Pu\| + C\|Pu_n - y_n\|\end{aligned}$$

Hence, we get for $C_1 > C$,

$$\|h_n\| \le C_1\|Pu\| \tag{2.28}$$

if η_n in (2.26) is sufficiently small and n is sufficiently large, satisfying relations (2.23) to (2.25). In order to prove relation (2.20), the following identity is used.

$$P\bar{u} - (1 - \varepsilon)Pu = P(u_n + \varepsilon h_n) - (1 - \varepsilon)Pu_n + (1 - \varepsilon)(Pu_n - Pu)$$

Now let us utilize the argument of Lemma 2.3 and the following identity:

$$\begin{aligned}&P(u_n + \varepsilon h_n) - Pu_n + \varepsilon Pu_n \\ &= \{P(u_n + \varepsilon h_n) - \tilde{P}(u_n + \varepsilon h_n)\} - (1 - \varepsilon)(Pu_n - \tilde{P}u_n) \\ &\quad + [\tilde{F}(u_n + \varepsilon h_n) - \tilde{F}u_n] + \varepsilon(Ah_n + y_n) \\ &\quad + \varepsilon(Pu_n - y_n) + \varepsilon(\tilde{P}u_n - Pu)\end{aligned}$$

Since $\tilde{P}u - Pu = Fu - \tilde{F}u$, we obtain the following estimate:

$$\begin{aligned}\|P(u_n + \varepsilon h_n) - (1 - \varepsilon)Pu_n\| &\le 2\eta_n + K\varepsilon\|h_n\| + \eta_n \\ &\le \varepsilon K\lambda^{-1}\|y_n\| + 3\eta_n \\ &\le \varepsilon K\eta^{-1}\|Pu_n\| + \varepsilon K\eta^{-1}\eta_n + 3\eta_n \\ &\le \varepsilon q'\|Pu_n\|/2\end{aligned}$$

if $K\lambda^{-1} < q'/2$ and $\varepsilon K\lambda^{-1}\eta_n + 3\eta_n \le \varepsilon(q'/2 - K\lambda^{-1})\|Pu_n\|$, where $\|Pu_n - y_n\| \le \eta_n$ and $\|Fu_n - \tilde{F}u_n\| \le \eta_n$, by virtue of (2.10) and (2.11) with $u = u_n$, $\eta = \eta_n$, and $h = h_n$, a solution of equation (2.27). Hence, relation (2.20) holds if n is sufficiently large and η_n from (2.26) is sufficiently small, and if condition (2.24) is satisfied. Relation (2.21) holds if we put $\bar{u} = u_n + \varepsilon h_n$ and n is sufficiently large, by virtue of (2.28). For relation $\bar{u} \in V_0$ to be satisfied, we choose a fixed $0 < \varepsilon \le \rho \le 1$ for which condition (2.22) holds and n so large that (2.23) is fulfilled, and use relation (2.28). Notice that the restriction involving $\rho \le 1$ in (2.11) is needed because $0 < \varepsilon \le \rho$ chosen above remains fixed while u_n is moving toward u, and we need to apply condition (2.11) to u_n. The proof of the lemma is complete. ■

We are now in a position to apply the method of contractor directions to prove the following theorem.

2.1 THEOREM In addition to the assumptions (A_1) to (A_2), suppose that the operator P: $V_1 \to L^1(G)$ is closed on V_1, the closure of $V_0 = D(A) \cap S(u_0,r)$, where P is given by (2.1). If the radius r is such that $r > \bar{r} = (1 - q)^{-1}\bar{C}\|Pu_0\| \exp(1 - q)$, $C < \bar{C}$, where $\bar{C}$ and $0 < q < 1$ are arbitrary and $C > \lambda^{-1}$ with $2K\lambda^{-1} < q' < q$, then equation (2.1) has a solution $u \in V_1$.

Proof: We construct a sequence $\{u_n\} \subset V_0$ and a sequence $\{\varepsilon_n\}$ with $0 < \varepsilon_n \le 1$ such that

$$\|Pu_{n+1} - (1 - \varepsilon_n)Pu_n\| \le q\varepsilon_n\|Pu_n\| \tag{2.29}$$

$$\|u_{n+1} - u_n\| \le \varepsilon_n \bar{C}\|Pu_n\| \tag{2.30}$$

as follows. Suppose that $u_0, u_1, \ldots, u_n$ and $\varepsilon_0, \varepsilon_1, \ldots, \varepsilon_{n-1}$ are known. Then it follows from Lemma 2.3 with $u = u_n$ that there exist $0 < \varepsilon_n \le 1$ and $\bar{u}_{n+1} = u_n + \bar{\varepsilon}_n h_n$ such that

$$\bar{u}_{n+1} \in V_0 \tag{2.31}$$

$$\|P\bar{u}_{n+1} - (1 - \bar{\varepsilon}_n)Pu_n\| \le q\varepsilon\|Pu_n\| \tag{2.32}$$

$$\|\bar{u}_{n+1} - u_n\| \le \bar{\varepsilon}_n \bar{C}\|Pu_n\| \tag{2.33}$$

Then we choose ε_n to be approximately (up to a constant factor, $0 < c < 1$) the largest number $\bar{\varepsilon}_n$ for which u_{n+1} exists and satisfies relations (2.31) to (2.33). Now it follows from Lemma 2.1, Appendix 1, that $\|u_n\| < \bar{r} < r$ for all n, and there exist u and y such that

$$u_n \to u \qquad \text{and} \qquad Pu_n \to y \qquad \text{as } n \to \infty$$

Since P is closed on V_1 by assumption, it follows that $y = Pu$. It also follows from Lemma 2.1, Appendix 1, that if $\{\varepsilon_n\}$ is not convergent to 0, then $Pu = 0$. Therefore, suppose that $Pu \ne 0$ and

$$\varepsilon_n \to 0 \qquad \text{as } n \to \infty \tag{2.34}$$

It results from Lemma 2.4 that there exist $0 < \varepsilon \le 1$ and $\bar{u} \in V_0$, which satisfy relations (2.20) and (2.21). Since

$$\|u_n - u\| \to 0 \qquad \text{and} \qquad \|Pu_n - Pu\| \to 0 \qquad \text{as } n \to \infty$$

and the sequence $\{\|Pu_n\|\}$ is decreasing, it follows that

$$\|P\bar{u} - (1 - \varepsilon)Pu_n\| \le q\varepsilon\|Pu_n\|$$

and

$$\|\bar{u} - u_n\| \le \varepsilon\bar{C}\|Pu_n\|$$

for all sufficiently large n. But the approximate maximality choice of ε_n implies that $\varepsilon_n \ge c\varepsilon$, where $0 < c < 1$, and this is a contradiction to (2.34), and the proof of the theorem is completed. ■

Now let us replace (A_1) and (A_2) with the following assumption.

(A_3) We assume that $F: V_1 \to L^1(G)$ is locally Lipschitz continuous in the following sense. There exists a constant K which has the following property: For arbitrary $u \in V_1$, the closure of V_0, and arbitrary $h \in D(A)$, there exists $0 < \varepsilon(u) \le 1$ such that

$$\|F(u + \varepsilon h) - F(u)\| \le K\varepsilon\|h\| \tag{2.35}$$

where $u + \varepsilon h \in V_1$.

Then Lemmas 2.3 and 2.4 should be replaced by the following.

2.5 LEMMA Let $u \in V_1$ be given and let q be an arbitrary number with $0 < q < 1$. Then there exist $h \in D(A)$ and $0 < \varepsilon < 1$ such that

$$\|P(u + \varepsilon h) - (1 - \varepsilon)Pu\| \le q\varepsilon\|Pu\| \tag{2.36}$$

$$\|h\| \le C\|Pu\| \tag{2.37}$$

$$\bar{u} \in V_1 \tag{2.38}$$

where $\bar{u} = u + \varepsilon h$, $K\lambda^{-1} < q$, and $C < \lambda^{-1}$.

Proof: We use the following identity for $Pu = Au + Fu$.

$$P(u + \varepsilon h) - Pu + \varepsilon Pu = F(u + \varepsilon h) - Fu + \varepsilon(Ah + y) + \varepsilon(Pu - y)$$

Hence, we obtain, by virtue of (2.35),

$$\|P(u + \varepsilon h) - (1 - \varepsilon)Pu\| \le K\varepsilon\|h\| + \varepsilon\eta \le \varepsilon K\lambda^{-1}\|y\| + \varepsilon\eta$$
$$\le \varepsilon K\lambda^{-1}\|Pu\| + \varepsilon K\lambda^{-1}\eta + \varepsilon\eta$$

where $y \in D(A)$ and $\|Pu - y\| \le \eta$, $h \in D(A)$ being the solution of equation (2.12). Hence, we obtain relation (2.36) if $K\lambda^{-1} < q$ and $K\lambda^{-1}\eta + \eta \le (q - K\lambda^{-1})\|Pu\|$. Relation (2.37) follows from (2.14). ∎

2.2 THEOREM Theorem 2.1 holds if assumptions (A_1) and (A_2) are replaced by assumption (A_3).

Proof: The proof is based on Lemma 2.5 and is a slight simplification of the proof of Theorem 2.1. ∎

The following remark can be useful in practical applications of Theorem 2.2.

2.1 REMARK By using Lemma 2.5 with $u = u_n$, we can construct a sequence $\{u_n\} \subset V_1$ such that

$$u_{n+1} = u_n + \varepsilon_n h_n \qquad 0 < \varepsilon_n \le 1$$

$$\|P(u_n + \varepsilon_n h_n) - (1 - \varepsilon_n) Pu_n\| \le q\varepsilon_n \|Pu_n\|$$

$$\|h_n\| \le C\|Pu_n\|$$

for all n. If

$$\sum_{n=0}^{\infty} \varepsilon_n = \infty$$

then $u = u_0 + \Sigma_{n=0}^{\infty} \varepsilon_n h_n$ is a solution of the equation $Pu = 0$, and $u \in V_1$.

Proof: The proof results from Lemma 2.1 ∎

Let us mention that in order to prove the uniqueness of the solution of the equation $Pu = 0$, a stronger local Lipschitz continuity of F is needed.

References

Agmon, S., Douglis, A., and L. Nirenberg. Estimates near the boundary for solutions of elliptic partial differential equations satisfying general boundary conditions, Part I, *Comm. Pure Appl. Math. 12* (1959), 623-727; Part II, *Comm. Pure Appl. Math. 17* (1964), 35-92.

Altman, M. *Contractors and Contractor Directions: Theory and Applications*. New York: Marcel Dekker, 1977.

Altman, M. An existence principle in nonlinear functional analysis, *J. Nonlinear Analysis 2* (1978a), 765-771.

Altman, M. An application of the method of contractor directions to nonlinear programming, *Numer. Funct. Anal. and Optimiz. 1* (1979), 647-663.

Altman, M. A strategy theory of solving equations, in Proceedings of the Special Session on Application of Functional Analysis to Numerical Analysis, *Lecture Notes in Mathematics*. New York: Springer-Verlag, 1979a.

Altman, M. Contractor directions and evolution equations, *J. Nonlinear Analysis 3* (1979b), 325-336.

Altman, M. Contractor directions and monotone operators, *J. Integral Equ. 1* (1979c), 17-33.

Altman, M. Iterative methods of contractor directions, *J. Nonlin. Analysis 4* (1980), 761-772.

Altman, M. Quasilinear evolution equations in nonreflexive Banach spaces, *J. Integral Equ. 3* (1981a), 153-164.

Altman, M. Quasilinear evolution equations in nonreflexive Banach spaces, I, *J. Nonlin. Anal. 5* (1981b), 411-421.

Altman, M. A simplified approach to linear evolution equations, *J. Integr. Equ. 4* (1982a), 265-270.

Altman, M. Nonlinear evolution equations in Banach spaces, *J. Integr. Equ. 4* (1982b), 307-322.

Altman, M. The method of contractor directions for nonlinear elliptic equations, *Nonlin. Anal. 6* (1982c), 167-174.

Altman, M. A new approach to quasi-linear evolution equations in nonreflexive Banach spaces, *Nonlin. Anal. 7* (1983), 1081-1088.

Altman, M. A general local existence theorem for quasi-linear evolution equations in nonreflexive Banach spaces, *J. Nonlin. Anal. 7* (1983a), 1463-1470.

Altman, M. Nonlinear equations of evolution and convex approximate linearization in Banach spaces, *J. Nonlin. Anal. 8* (1984a), 457-470.

Altman, M. Quasilinear evolution equations and convex approximate linearization in Banach spaces, *J. Nonlin. Anal. 8* (1984b), 471-480.

Altman, M. Nonlinear evolution equations and smoothing operators in Banach spaces, *J. Nonlinear Anal. 8* (1984c), 481-490.

Altman, M. Nonlinear equations of evolution in Banach spaces, *J. Nonlin. Anal. 8* (1984d), 491-499.

Altman, M. A theory of nonlinear evolution equations, *North-Holland Math. Stud. 110* (1985a), 17-26.

Altman, M. A theory of nonlinear operator equations based on global linearization iterative methods in Banach spaces, International Conference on the Theory and Applications of Differential Equations, May 20-23, 1985b, Pan Amer. University, Edinburg, Texas.

Altman, M. Strongly quasilinear evolution equations in Banach spaces, *J. Intgr. Eq. 9* (1985c), 95-112.

Altman, M. Nonlinear evolution equations via smoothing operators combined with elliptic regularization, *J. Intgr. Eq. 9* (1985d), 179-197.

Altman, M. Global linearization iterative methods for nonlinear evolution equations I, II, *J. Intgr. Eq. 9* (1985e), 125-134, 169-177.

Altman, M. Global linearization iterative methods and nonlinear partial differential equations, I, II, III, to appear.

Altman, M. A new approach to approximate linearization for nonlinear evolution equations, I (1986), to appear.

Barbu, V. *Nonlinear Semigroups and Differential Equations in Banach Spaces.* Editura Acad. Romania, 1976.

Bondarenko, V. A., and Zabreiko, P. P. On a theorem of J. Moser, *Vestnik Iaroslav. Univ. 8* (1974), 3-7.

Case, K. M., and Zweifel, P. F. *Linear Transport Theory.* Boston: Addison-Wesley, 1967.

Cesari, L. The implicit function theorem in functional analysis, *Duke Math. Jour. 33* (1966), 417-470.

Cesari, L. Functional analysis, nonlinear differential equations, and the alternative method, in *Lecture Notes in Pure and Appl. Math. 19* (1976), 1-197. New York: Marcel Dekker.

Dorroh, J. R., and Graff, R. A. Integral equations in Banach spaces, a general approach to the linear Cauchy problem, and an application to the nonlinear problem, *J. Integral Equations 1* (1979), 309-359.

Friedrichs, K. O. Symmetric hyperbolic linear differential equations, *Comm. Pure Appl. Math. 7* (1954), 345-392.

Graff, R. A. A functional analytic approach to existence and uniqueness of solutions to some nonlinear Cauchy problems (1979) (preprint), LSU.

Hazan, M. I. Nonlinear evolution equations in locally convex spaces, *Dokl. Akad. Nauk SSSR 212* (1973), No. 6; *Soviet Math. Dokl. 14* (1973), No. 5, 1608-1613.

Hazan, M. I. Existence and approximation of solutions of nonlinear evolution equations, *Latviisk. Mat. Ezhegod. 26* (1982), 114-131 (Russian).

Hazan, M. I. Nonlinear and quasilinear evolution equations: Existence, uniqueness, and comparison of solutions: The rate of convergence of the difference method, *Zapiski Nauc. Seminar. LOMI, Tom 127* (1983), 181-200.

Hörmander, L. The boundary value problems of physical geodesy, *Arch. Ration. Mech. Anal. 62* (1976), 1-52.

Hughes, T., Kato, T., and Marsden, J. Well-posed quasi-linear second-order hyperbolic systems with applications to nonlinear elastodynamics and general relativity, *Arch. Rat. Mech. Anal. 63* (1977), 273-294.

Kato, T. Linear evolution equations of "hyperbolic" type, *J. Fac. Sci. Univ. Tokyo, I, 17* (1970), 241-258.

Kato, T. Linear evolution equations of "hyperbolic" type II, *J. Math. Soc. Japan 25* (1973), 648-666.

Kato, T. Quasi-linear equations of evolution with applications to partial differential equations, in *Lecture Notes in Math. 448*, pp. 25-70. New York: Springer-Verlag, 1975.

Kato, T. Quasi-linear equations of evolution in nonreflexive Banach spaces, Lecture Notes in Num. Appl. Anal. *5* (1982), 61-76. Nonlinear PDE in Applied Science, U.S.-Japan Seminar, Tokyo, 1982.

Kobayashi, K. On difference approximation of time dependent nonlinear evolution equations in Banach spaces, *Memoirs of Sagami Inst. Tech. 17* (1983), 59-69.

Krasnosel'skii, M. A., and Rutickii, Ya. B. Some approximate methods of solving nonlinear operator equations based on linearization, *Soviet Math. Doklady 2* (1961), No. 6.

Krugliak, N. Ia. A modification of a theorem of J. Moser on solvability of nonlinear equations, *Vestnik Jaroslav. Univ. 12* (1975), 77-86.

Moser, J. A rapidly convergent iteration method and nonlinear partial differential equations, *Ann. Scuola Norm. Sup. Pisa 20* (1966), 265-315.

Nakata, M. Quasi-linear evolution equations in nonreflexive Banach spaces and applications to hyperbolic systems, Thesis, Univ. of California, Berkeley, 1983.

Nash, J. The embedding problem for Riemannian manifolds, *Ann. Math. 63* (1956), 20-63.

Pao, C. V. Solution of a nonlinear Boltzmann equation for neutron transport in L^1 space, *Arch. Rat. Mech. Anal. 50* (1973), 290-302.

Petzeltova, H. Application of Moser's method to a certain type of evolution equations, *Czechoslovak Math. J. 33* (108) (1983), 427-434.

Rabinowitz, P. H. Periodic solutions of nonlinear hyperbolic partial differential equations, II, *Comm. Pure Appl. Math. 22* (1969), 15-39.

Schwartz, J. T. *Nonlinear Functional Analysis, Courant Inst. Math. Sci.* New York: New York University, 1965.

Sergeraert, F. Une generalisation de theoreme des fonctions implicites de Nash, *R. C. Ac. Sci., Paris 270* (1970), Ser. A, 861-863.

Tanabe, H. *Equations of Evolution*, Pitman, 1979.

Yosida, K. *Functional Analysis*, 2nd ed. New York: Springer-Verlag, 1968.

Index